FUSION
NOW!

TOMORROW'S ENERGY
TODAY

A YOUNG TEAM OF SCIENTISTS STRUGGLE TO SOLVE THE ENERGY PUZZLE . . .

A NOVEL BY: W.E. POWELSON

Copyright

Table of Contents

Acknowledgements

Thanks again to Jenny A. Anderson, McMinnville, Tennessee for the editing, proofreading, advice and help.

Introduction

It's late Fall in the year 2016. Two drinking buddies; one a Lieutenant General at the DOD and the other a civilian business man, discuss the advantages of developing fusion energy. The obvious conclusion is that the government needs to create a **second Manhattan Project.**

Both men realize that the government could do in 4 years or less, what the private sector had failed to do in more than 60 years of empty promises.

Fusion energy promised to solve 80% of the World's problems quickly and in one swoop, if only someone in the right position was to take the initiative.

Global Warming, Terrorism and poverty are just a quick sampling of problems that would easily be solved if sustained Fusion was available.

It turns out that our 3-star General in the story is the one man in the World to get it done. General Frank Wilson just happens to be the General in charge of Research and Development deep inside the Department of Defense.

Both men are aware that once fusion arrives, all the other forms of energy will immediately become obsolete. There's no doubt or debate about any of that. Fusion is the ultimate energy because, unlike Fission, it promises **ECSAFE: E**conomical, **C**lean, **S**afe, **A**bundant, and (virtually) **F**ree **E**nergy. Fusion energy had been possible since the 1950's. The problem was . . . containment.

Our two heroes launch a quest to develop fusion containment technology **now . . .** rather than later. Because they are well aware of how desperately we needed it yesterday or sooner. They also know that later . . . may be too late!

It's so obvious! Fusion energy would end the Terror Wars and it would quell all the fears of Global Warming. The economy of the entire World would surge to unprecedented levels. A new Manhattan Project; developing fusion energy would offer the **one solution** and instantly bring the World back from the brink!

The General becomes an icon of history because he is in the precise position to acquire the funding as a **black budget project** and he alone holds the power to get the job done.

As a result of some gentle prodding by his civilian friend, he decides to do it, but then he surprises his friend and chooses him, (the civilian business man) to become the director of the project. They name the project, "**Project Starshine**".

Together, with the might of America behind them, they locate a team of four MIT grads with Ph.D's in experimental plasma physics, to do the scientific and experimental work at Los Alamos, New Mexico. The special team is chosen because they have different and unique, new ideas about how to contain the heat extremes inherent with fusion technology. They are a team of especially gifted out-of-the-box thinkers, with can-do attitudes. "**Can't**" and "**impossible**" are two words they simply do not recognize.

They know they can accomplish the task and they are determined to succeed in record time, and well under budget.

Add a couple of beautiful women with genius level minds. Mix them gently with equally interesting male intellects . . . and stir. A story of love, greed, intrigue, espionage and determination will explode within your heart, mind and soul.

Will they do it?

Will they succeed?

How?

Read on, to discover for yourself just how easily it might be done, and what it would mean.

What if our government might actually wake up and get fully behind such a project as if: failure was not an option?

Because it truly isn't!

*Probability math will prove that most options are possible,
given enough time-tested quirks of random fate . . .*

Chapter 1: Storm Clouds Approach.

The Friday afternoon 5:30 PM sun peaked from behind a mass of gray overcast, amid calm breezes. From the East, a darker more ominous and threatening bank of black clouds approached at a rapid pace. Palm Beach Florida braced for a late season level-three Fall hurricane.

The tires of a private Learjet 45XR burped twice as they touched down smoothly on the runway at the West Palm Beach International Airport. The jet taxied to a stop near a stretch limo waiting at the end of the tarmac.

One lone occupant stepped off the the plane as a chauffeur greeted the new arrival and escorted him to the waiting limo. He opened the rear door with a gracious smile as the quiet man took a seat inside the luxurious and spacious vehicle.

The smartly dressed chauffeur then quickly retrieved and carried a large suitcase and a briefcase from the plane. He hurriedly carried it all to the rear of the limo and loaded the luggage into the trunk. At a near run, he then took his place behind the wheel, buckled up, and simultaneously pushed the intercom button,

"Shall I drive directly to the estate, sir?"

"Yes, the estate. Thank you," came the relaxed answer.

As the stretch limo arrived at an ocean front mansion, the automatic gate quickly opened without hesitation, and the limo crawled the circular driveway to a stop at the front entrance of the expansive two-story Spanish Colonial mansion. The passenger noticed that the 40 acre lawn and landscape were immaculately manicured. He made a mental note to compliment the yard keepers for their exemplary work.

The lone occupant in the rear of the limo, known only as Mr. X, may quietly be one of the most powerful and important men in the World. But; he is the type of man who prefers to shy away from the spotlights. Very few people are aware of just how important he actually is, and that's the way he prefers it.

As the chauffeur leaped to open the rider's door, the nameless Mr. X emerged from the immaculate vehicle and leisurely approached the 12 foot hand-carved double doors of the opulent residence. The chauffeur was close on his heels. Charles the chauffeur swung open the door, and the mysterious man entered. The house was totally empty except for the two men.

"Please carry my luggage to the upstairs master bedroom Charles; then be certain the limo is positively safe and secure from the approaching storm. Do the same for yourself as well. That will be all for today. Thank you."

Mr. X was relatively young, in his mid-40s and very distinguished. He was in great physical shape. His hair was sandy blonde, neatly trimmed, but intentionally a little long and over the ears. He wore a polo pullover and dress pants with expensive black Italian loafers that glistened. He seemed to constantly wear a gregarious, friendly, welcoming smile, every minute of every day. The self-made multi billionaire was 46 years old and still single. In his own mind he felt that he had just never met the right woman. His age, appearance and status caused women in every age group to stop, ponder and fantasize. He had made his fortunes in the hi-tech boom of the late 1990s, then invested wisely. Mr. X was one of the rare people on the planet who made more money every day than he could spend in a lifetime.

As the chauffeur returned from the upstairs bedroom, Mr. X smiled and handed him two crisp $100 dollar bills as a tip, along with a friendly pat on the back. "Treat your wife to something nice, Charley. See you after the storm. I'll call when I know my itinerary. Have a safe and enjoyable weekend."

The chauffeur disappeared through the 12-foot hand-carved double entrance doors and was gone. The house fell silent for a few minutes until Mr. X pushed the buttons on the TV remote and turned to channel 3 for the latest weather update.

The mansion is Mr. X's vacation home, and he has arrived specifically because a level three hurricane is churning and twisting it's way in from the southeast Atlantic ocean on a direct collision course with Palm Beach. Mr. X simply wants to experience the ferocity of the storm first-hand from his second-floor bedroom balcony overlooking the Atlantic ocean.

While most people evacuate during storms of this magnitude, this man enjoys them. It's a mark of the man and his personality.

He simply loves the awesome power of nature and it's ability to challenge even the most powerful of men . . . himself included.

A radar map appeared on the 58-inch flat screen television as the local weatherman explained the predicted path and course of the dangerous storm that had been labeled Monique . . .

"According to the NOAH forecasters in Miami beach we here in West Palm Beach should expect the early rain-bands and first-squall lines to begin approaching the beaches around 7:30 PM.

As fast moving Monique approaches from the southeast, the winds will continually increase until around midnight at which time the eye of the storm should be directly overhead. We are expecting peak wind speeds of 110 to 120 miles per hour as the eye-wall passes our area. After a 20 to 30-minute lull, the winds will shift from the north and subsequently begin blowing from the west and southwest. The most forceful winds are always nearest the eye-wall. After the eye of the storm passes overhead, the winds will begin again in the 90 to 120 mile per hour range at around 12:30 to 1:00 AM, then taper off as the storm passes on to the northwest. We expect it to pass beyond West Palm Beach by 3:00 AM to 4:00 AM, barring any unforeseen changes in the conditions propelling the storm. The local fire and police . . . "

Mr. X turned the tv off and strolled upstairs to the master bedroom with a bottle of Jim Beam and enough ice and coke to last several hours as the hallway clock chimed 7:00 PM. He mixed a healthy drink and pulled the curtains in the bedroom full open, revealing the massive expanse of the Atlantic ocean to the east. The whitecaps were in full bloom all the way to the horizon and the storm surge was showing on the beach below about 200 yards to the east. A light rain band began peppering the area as the breezes intensified and gusted momentarily then settled again. The bedroom balcony was dry, so Mr. X took a seat in a lounge chair and admired the spectacle. There was little to worry about. The sliding glass doors of the bedroom leading to the balcony were certified in winds up to 175 miles per hour. This storm probably wouldn't go beyond 120 miles per hour. At the moment, the cool, strong breezes felt refreshing on his face and skin as he took a long, deep breath of the damp, salty air. It was exhilarating, and that was exactly what had lured him here.

Mr. X had traveled all the way from his permanent home in Washington D.C. just to experience the storm. The forecasters had been watching and predicting a direct hit to the West Palm Beach area for nearly a week. By early morning on Friday it appeared to be certain that the storm would definitely hit. He had alerted his pilot and crew and then made arrangements to spend the weekend in his Palm Beach mansion, specifically to enjoy the freak of nature that was being called Monique.

He had a lot on his mind. The intoxicating environment and solitude would help him concentrate and focus directly on the job at hand.

Chapter 2: TGIF at the Pentagon

5:30 PM Friday, Nov. 4, 2016: Pentagon / DOD.

Eddy G. Feywisch was one of the chief computer specialists at the Pentagon. Eddy's history was squeaky clean. He had served valiantly in Iraq and was an honored veteran. His friends high inside the Pentagon had helped secure the job for him but he was hired because he was the best choice. There had been background checks all the way to his childhood in Ocala Florida.

The job was long-term and important, as far as computer specialists jobs were concerned. No other computer job in the nation was as closely scrutinized as the crew that maintained and operated the systems inside the Pentagon. Eddy's job put him inside the Intelligence headquarters of all branches of the military. Those computers were also directly linked to the head-offices of the CIA, the FBI and the NSA. His job was to maintain the security and integrity of all the computers in the system. It was a complete nightmare of near inhuman responsibilities, but he was chosen for his ability to handle the job. Eddy was a 'can-do' sort of guy.

If there were other more important computer jobs anywhere in the World, it would only be the systems of the CIA, NSA and FBI. The NSA job coincidentally was the responsibility of his girlfriend Rachel Dunkin. She had top level security clearances as well. Her job was to supervise the security and integrity from her post at the NSA headquarters a few miles away. Between them, the two lovers had access to the government's most sensitive and delicate secrets. It was their job to keep those sensitive secrets secure and safe. They both took their jobs very seriously.

Eddy and Rachel were both doing quite well. Their combined salaries totaled over 210,000.00 per year. They were planning to get married soon. It was a matter of waiting for just the right moment in time. First, they would need to find a simultaneous break in each of their hectic schedules. That was an impossible dream that fluttered unrealistically somewhere in the far away future. They could only imagine it on the best of days.

They had met quite by accident as they shared a cab one night at the end of a very long day. Eddy had noticed that Rachel was wearing the familiar uniform of a NSA Intelligence Systems Analyst, as they both entered a cab at the same time. Since they were both going the same direction they decided to share the cab fare. A conversation had ensued and they both became entranced at the coincidences of their lives. She too had served in Iraq and had gone through all the top level security clearance evaluations and scrutiny.

They had a lot in common and decided to have a quick drink together as they discovered that they only lived two blocks apart on the same street. There was a neighborhood bar exactly in-between their two homes, so they began a routine of meeting there every evening as a way to wind down. One thing led to another until they had finally fallen in love and become a couple.

On this particular Friday, Eddy hurried himself through the finishing details of a system-wide security check.

The work load was light for Friday as most of the Pentagon staff had taken the day off in order to enjoy a three day week-end. Eddy's job mainly consisted of running simultaneous virus and integrity checks and searching for breach attempts. He barreled through the tasks quicker than usual. All the computer terminals appeared to be clean.

Inside the DOD main office he went through the usual routine doing the security and integrity checks, but as he worked at the setups, a secretive and seemingly sensitive e-memo on the computer popped-up unexpectedly and caught his eye. It appeared to be a simple office memo, but it was somewhat suspicious. He had already read it before he could stop himself. It was his job to be aware of anomalous curiosities. He was only being thorough.

FOR YOUR EYES ONLY!
Lieutenant General Frank G. Wilson:
Advanced Research Projects of the
Army, Navy, Marines and Air Force
at the Department of Defense, Pentagon:

"Frank: *I will be in West Palm for a few days and expect to return on Monday. Our plan is progressing nicely We must inform the new President Mrs. Clayton of our new project objectives.*

She needs to be fully advised that she will benefit more than anyone if we are successful. The accolades will fall directly on her and her administration as we quietly do the job for her, behind the scenes. Be certain that she fully understands the importance of Operation Starshine'. Sustained fusion is the future of the

World. The first country to develop it will dominate. Explain it to her in detail if necessary.

We are moving into phase-two. We are ready to put our team in place and begin in earnest. It is mandatory that the operation must remain top secret, black-budget and there are no alternatives to that.

Use your influence to keep the press out of this one, Frank. It needs to be approved very quietly and we want her to be on-board with us completely. Nothing since the Manhattan Project has ever been any more sensitive or important to the United States.

We both know we're taking-on a near impossible dream. If it remains secret and we fail, the secrecy will help us (and her) save face. We fully intend to succeed. Handle our new President with kid-gloves and try to gain her cooperation without threats. We would hope that she will be as enthusiastic as we are concerning this project. If she seems to stubbornly balk we may resort to pressures, the old way, as needed.

Finally, convince her that if we do succeed she will get the Lion's share of credit and the entire World will benefit greatly. This project could make her a Saint in the eyes of the World. She may become known as the greatest President in U.S. history, if we succeed. It will end the energy crisis, it will end global terrorism, and it will reduce the dangerous daily Carbon levels we're putting into the atmosphere, by 85%. So, failure is not an option! Should we fail, the project will remain top-secret forever. No one will ever know. This is a win/win for her and the entire World as well. " The note was signed . . . Mr. X.

Eddy was angry at himself for scanning the secret and personal memo. It could be considered a breach of security if anyone but him had seen it. Curiosity killed the cat! He tried to forget what he had just seen, but it was too late for that. Just the same, it was intriguing and interesting.

'Project Starshine': the sustained Fusion project, was on his mind as he finished his duties and prepared to go home.

By 2016, everyone knew that if scientists could find a way to contain the heat extremes inside nuclear fusion reactors, it would solve 90% of the World's problems.

As opposed to the current Fission reactors now melting down in places around the world; Fusion had been deemed **ECSAFE**: **E**conomical, **C**lean, **S**afe, **A**bundant, **F**ree Energy.

Then, he grinned. He was especially aware of a certain lady (his sister) who would dearly love to be aware of Project Starshine and the mysterious Mr.X. Yet, this was something he should ever mention or talk about with anyone. He considered it a lucky thing that he had seen the memo, and no one else. As he re-encrypted the memo, he wondered who the heck this mysterious Mr. X might be. He had a few clues and thought he knew, but he wasn't 100% certain.

"My sister's future may reside in that little memo." he said under his breath. "Someone needs to tip her off. I have to do it!"

Chapter 3: Pentagon Briefing for President Elect

The campaign trail had been especially rough in 2016. It was all about dredging up bones and slinging mud. The general public was appalled by it all. Voter turnout had been dismal but Henrietta Clayton had won by a fairly wide margin over Jeff Brush. Poor Jeff was following in brother George's shoes and no one wanted a repeat of that. At this point in history, the average American was totally and completely burned and confused by their embarrassingly inept government. Politics as usual meant lies and more lies, as usual.

Yet Henrietta Clayton, a Democrat, had emerged as the winner. She had carried herself well through all the debates and only resorted to mudslinging a couple of times. Everyone was fully aware that she had actually been the brains behind her husband Bob Clayton's very successful two terms as President, a few years earlier in the 1990s. The United States was hoping for a repeat of those excellent years.

She had won because she was not only the most intelligent contender, but she actually seemed to sincerely want the best for the country, in spite of what the either party might want. She offered the most hope to the people. Consequently she was elected by a comfortable margin.

Republicans were pounding the war-drums again. Iran was aspiring to become a military equal with all her nearby threats . . . Israel, India and Pakistan. Her neighbors had the bomb and she didn't. A tentative peace had been brokered in 2015, but it was a fragile peace at best and the Republicans had already vowed to undermine the treaty efforts.

Henrietta Clayton appeared to be totally against allowing any of that. It had won her a lot of votes. The average intelligent American knew that another senseless war over oil would only bring more disaster to their lives. They had already seen enough.

The World's economy had been in absolute shambles since 2008, as a direct result of the terror wars. The wars were for oil, though no one in politics or in the press would ever admit it openly. It wasn't easy for them to make the terrorists appear to be drooling maniacs, but they managed to do a fairly good job of it.

The American people would believe almost anything after the horrendous attacks on 9/11/01.

All but the largest banks had failed as a result of expensive Wars and disgusting, criminal management by Wall Street and the Federal Reserve. The too big to fail banks had then been propped-up precariously by the U.S. government and the FED, thanks to American taxpayers. The U.S. had been teetering on monetary collapse ever since.

The U.S. unemployment average was nearing 10% and shanty-towns for the homeless had sprung up in wooded areas all across the United States. The unemployment ratios were probably much higher, but 90% of the unemployed were refused unemployment payments as a way to keep the ratios superficially low as far as the news media was concerned. The true unemployment ratios of 20% would look bad and might cause further downturns in the economy. The solution to that was easy. All they had to do was refuse unemployment payments to the unemployed. They made the funds so difficult to obtain that even deserving people wouldn't endure the emotional and trying ordeal of applying for their rightful funds. No one offered any real solutions.

The Republicans blamed the Democrats and the Democrats blamed the Republicans. Congress sat on its hands as always, discussing, arguing and ignoring budget decisions concerning the plight of their starving, impoverished nation. Most agreed that the New Depression was in many ways worse than the Great Depression of the 1930's. The only difference between then and now was the fact that the impoverished of 2009 all seemed to enjoy cable TV, cell phones and Internet connections as they starved in their foreclosed homes.

President Henrietta Clayton picked up the ringing iPhone. "Henrietta Clayton here." She responded, "May I ask who is calling?"

"Good morning Ms. President. I am Major Jeremy Smith, but that isn't important. I'm only the messenger. The War Department will be having a routine meeting, scheduled to occur at 13:00 on Monday at the Pentagon. Your attendance is humbly requested. We congratulate you on your recent win in the election and we look forward to serving you for the next four years. The War Department wants to discuss a few new directives for the coming term. We'll tend to work hand-in-hand with you for the next four years, Ms. President. As you know, this briefing is something we do just shortly after each election. We'll be wanting to brief you on all the major affairs foreign and domestic, requiring your approval and attention. Will you be available to join us at the DOD over at the Pentagon on Monday at 13:00?"

"Yes Jeremy. I can be there. I suppose I'll see you then."

"Okay fine and Ms. Clayton, please come alone. What we must discuss is for your ears only. Please, no entourage, the matters we'll be discussing will all be of a top-secret nature, for your ears only. No one but you should be privy to the information that will be discussed."

"Yes, of course. But, the secret service goes everywhere I go. You know that! What about them? I don't think I can shake them, even if I try. "

"They'll be taken care of for you. We can control the Secret Service from here. They will escort you over here, but they'll back away once you enter our office. They'll get a little doughnut break. No problem. See you on Monday then. We're looking forward to meeting with you ma'am. We have much to discuss. The future of the World is at stake. We have plans that will surely prove to help make you one of the most popular Presidents in U.S. History. We're hoping that we can work as a team."

"Okay, Monday at 13:00 it is. See you then." She ended the connection feeling slightly threatened but she wasn't certain why.

Chapter 4: Monster Monique.

9:30 PM, Palm Beach Florida:

By 9:30 PM in Palm Beach Florida, Monique was in full bloom. The rains were blowing horizontal as gusts of wind lashed and slammed the sides of the Mansion.

Mr. X had finally retreated back inside the bedroom. He sat with the lights off, behind the glass sliding doors, quietly sipping a Jim Beam and Coke as sheets of rain peppered the ocean side of the building. Occasional glimpses of the ocean out front revealed towering waves of 10 to 20 feet that now broke over the seawall and washed across the street below the mansion. Palm fronds flew in every direction as the tall palms bent almost double from the force of the wind.

A small metal roadway sign from somewhere blew down the street about 100 yards in front of the mansion and slammed into a parked car on the street below, shattering the windshield. It then became airborne again and flew over the roof of the car. It continued on down the street and finally became lodged in the shrubbery of a neighbor's yard. Suddenly a loud, crashing thump resounded above the howling winds as an old Sycamore tree fell harmlessly to the ground. It had become completely weakened by the water soaked earth and uprooted by the winds.

Mr. X's cell phone rang in the darkness. It was a young Captain Dugger, from General Frank Wilson's staff. He had been designated as a gofer or helper to assist Mr. X with the formation of the proposed 'Project Starshine'.

Mr. X answered, "Hello."

A voice on the other end responded,"Sir, I thought it might be good to give you a heads-up on our progress. Things are moving along rapidly.

General Wilson has assigned me with the task to help you locate prospective physicists for Project Starshine. He also wants me to look for dirt on President Hennrietta Clayton. I swear, the old gal is squeaky clean. We can't find anything in her past that we can use. I don't believe she has ever done anything questionable that the entire country doesn't already know about. Even then, she has always been vindicated. We're working on it! I have three men on the detail and two women. Something is bound to turn-up sooner or later, but so far, this gal is cleaner than a Sunday school teacher.

Our Project Starshine team may be shaping-up nicely. I've sent you an e-mail-attachment. It contains the dossiers of ten physicists we think may qualify for selection as the scientific team for the project. We have a green light to use the facilities at Los Alamos and we have also secured the testing facilities at nearby Sandia in Albuquerque too.

Man! What's all that racket? It sounds like you are in the middle of a hurricane!"

"I am. I'm in Palm Beach right now. That's Monique you are hearing. She's quite a violent little lady . . .

Listen! Don't worry too much about getting the goods on President Clayton. I'm not a fan of those tactics. I don't think she will be a problem. The Starshine Project should stand on its own merit. It's a terrific idea, plain and simple. The new President will quickly see that its a win/win for her and nearly everyone in the country, (except the big oil companies). If she isn't catering to them, I'm certain she will approve the budgeting money. If she refuses, we'll worry about pressuring her as the needs arise. I'd say, keep looking, but don't worry if she is really clean. Maybe it's a good thing! Maybe she's actually a worthy President! That would be nice switch, wouldn't it?" He was being a little tongue-in-cheek, sarcastic.

"Will do boss! Enjoy your hurricane and pour a JB & Coke for me." They ended the connection.

Mr. X was indeed enjoying the hurricane. By 11:30 PM the eye-wall was approaching. The violent winds were sustained at 115 mph with occasional gusts at around 130 mph. The glass sliding doors that overlooked the balcony and the Atlantic ocean rattled constantly now. The thunderous roar of the wind offered a consistent subtle reminder of nature's awesome power. Something about the power of the vortex winds made Mr. X think of Fusion confinement. He wondered if a gravitational vortex configuration had ever been tried to help contain the resulting heat of a fusion reaction. A palm frond flew through the air and landed squarely on the glass with a loud whop.

Mr. X was by now enjoying quite a buzz from the Jim Beam and Coke. The palm frond startled him. It was loud and the force of the wind held it in place for several seconds before it shifted slightly, which sent it sailing off and away from the balcony. It soon disappeared into the nearly horizontal sheets of rain and black of night.

Then came the calm. By midnight, the calm inner-eye of the storm was overhead. Everything stopped and it was suddenly very quiet. The rain and the wind had stopped completely. The Atlantic was still churning wildly but the clouds cleared completely and a starlit sky gleamed overhead for nearly 30 full minutes.

Then the second half of the storm came on quickly and strong by 12:40. The winds weren't quite as strong after the wind directions shifted, but the second half of the storm continued. By 1:20 AM most of the excitement was over. The storm tapered off from there and by 3:00 AM it had moved on towards Orlando.

Mr. X stepped onto the balcony again and felt the stiff breeze on his face. The breeze was now barely 25 mph and the driving rain had turned to a fine misty drizzle.

He finished his final JB and Coke on the balcony, then went to bed. Saturday would be a day to relax and then on Sunday evening he would fly back to DC.

On his laptop he found the email attachment Captain Dugger had prepared and sent. It contained the names and contact info of what might become the most elite scientific team of the 21st Century.

In the early 1940's of the 20th Century, the Manhattan Project had fostered household names like Groves, Oppenheimer and Feynman. He wondered whose names might reign in the 21st Century. It was his job to locate, qualify, then choose that soon to be, very elite group of physicists.

Chapter 5: Meeting With The DOD.

At 12:45 Monday, just after a quick and worried lunch, the new President, Ms. Hennrietta Clayton boarded the presidential chopper for a quick ride to the Pentagon. She was accompanied by an entourage of Secret Service personnel. A few minutes later they touched down on the chopper pad at the Pentagon and she was escorted inside the building and to the main office of the DOD. As she entered the meeting room, all but one of the security agents headed for the cafeteria. One agent remained posted outside at the door of the meeting room.

She was immediately greeted by two Pentagon guards in snappy Marine and Air Force uniforms. Each of their uniforms displayed a string of medals down the front. The two men were obviously distinguished war veterans, now being employed and very well paid, as escorts. She was ushered to a long oval table where a cadre of military brass and Department of Defense officials were already seated. They were all talking, sipping drinks and a few were vaping their smokes.

Electronic Vapes we're being received everywhere as the new more acceptable form of smoking. Try as they might; the rabid control-freaks had not found enough evidence that vaping was harmful or unhealthy and dangerous. In fact, the scientists at Johns Hopkins had recently reported that nicotine had proved to be an important deterrent to Alzheimer disease! New reports were proving that habitual smokers were far less susceptible to the ravages of the dreaded disease. The researchers were now injecting Alzheimer patients with nicotine and achieving astounding results. Some of the demented patients were actually regaining their old memories. The FDA had struggled to suppress the new knowledge and keep it under wraps, but the smokers had made a big deal out of the findings. Vaping was in! Technology had made America a free country once again, (at least temporarily).

President elect Clayton was seated at the center of the long table as a waiter took her drink order. She asked for sweet tea.

The meeting came to order. Seated at the long oval table with her were all the top military leaders at the Pentagon. Many of the men at the table wore two or three stars on their epaulets and each proudly displayed lengthy arrays of medals on the left breast of their uniforms. Some were civilian DOD officials.

Finally General Frank Wilson spoke. "Madam President, I fear that this meeting may begin to sound a little bit like a high pressure sales pitch. I hope it doesn't, but the matter we wish to discuss deserves a full explanation. We are hoping to make some serious history here today and you will become a large part of it.

Of course we all know the history of the Manhattan Project that took place at the beginning of World War II. At the time, it too was a top secret shot-in-the-dark that began with a hope and a prayer. The Manhattan Project eventually proved to be a major game changer for the entire World and the United States in particular. Our government has been riding high on that accomplishment ever since. It helped us to become the strongest and most advanced country in the history of the World. The Atomic Age was born from the Manhattan Project.

The interesting point here is, as a result of the Manhattan Project, our government was able to accomplish in 4-years what might have taken the private sector 50 years to accomplish. As a consequence, we ended the War and became the most powerful nation on earth.

Of course, at that time, in 1942, we were under tremendous pressures and threatened with complete annihilation from both Germany and Japan. Our very existence was threatened at the time. We simply had to succeed! Failure was not an option. Our agenda today is just as hopeful as that one was, but not quite as threatening. Or is it . . .?

We here at the Pentagon have become aware that a similar project begs to be undertaken at the present time. We should not wait until the knives are at our throat, to act. It is in the interest of preparedness that we have evolved an idea that promises to end the Terror Wars completely. It will end the Energy crisis and solve the ecological CO_2 (Global Warming) equation in one fell swoop. If we are successful, the world's economy will soar to levels we've never seen before.

We have in mind a proposition for a new project that, if successful, will grant you a lasting place in history as possibly the greatest President in the history of this nation. We will quietly do the difficult work behind the scenes for you, but you and your administration will reap the accolades and rewards when and if we are successful. We feel that you are the right President at the right time to benefit from this immense task we have in mind. Do I have your attention at this point?"

President Clayton calmly took a sip of tea and responded somewhat casually, without blinking, "Well yes! I'm quite anxious to know what you are driving at, General. Please continue." She smiled graciously but with the radiant wisdom that had placed her into the position she now held. 'Her smile spoke volumes without words, *"This woman is open-minded but certainly not a push-over."*

Hennrietta realized that she was a woman in the seat of great power, but she was also up against some of the most highly trained mental giants and power-hungry men our military academies and universities had ever produced. In a way, she was one against them all. Most people would have been terrified and completely intimidated to be in such a situation. She had been around DC most of her life and career. She was adept at handling this, and nearly any situation. She remained cool but attentive.

"Well, Ms. President; what we're talking about here is a black-budget idea that we have come-up with. It won't come cheap, but it may offer the highest ever, possible return on investment. When compared to all the other ideas that we've undertaken since the famed Manhattan Project, it dwarfs them all, including the Manhattan Project and the more recent Stealth Project. It is very similar in nature to the Manhattan Project. It is primarily an extension of it. At the current time we are referring to it as the 'Starshine Project'.

We've listened to the physicists and scientists for years. They have always promised that Fusion containment would arrive in 10-years. We've been hearing them say that for the past 60 years. We feel that their 10-years is up. It is time to produce! We want to do it in 4-years or less! It is the end-all solution to nearly every major problem that plagues our societies today.

It is a known scientific fact that Fusion energy will be the most efficient, most economical energy solution we have the means to conquer. It will mean that the World will soon go 100% electric, as maybe it should have done 100 years ago. But the new energy will be environmentally safe. We'll cut carbon emissions by more than 85% or 90% as soon as Fusion electric goes mainstream. We will no longer need the Arab oil reserves. We can pull all our bases out of those regions. Terrorism will quietly fade into history.

Essentially, the economy will rebound to the 1995 levels we enjoyed when your husband was president. The economy may surpass even those levels, because fuel costs will flatten all the way back to 1960 levels. A person will be able to drive completely across the United States on less than $5 worth of fuel. It will decrease the cost of transporting goods around the World. The markets will all go berserk and into the green. With the exception of the fossil fuel industries, of course.

We want to put together a team of 'can-do' outside-the-box thinkers and scientists to go to work on the fusion containment technology, and finally conquer it. It will keep America strong and it will bring lasting peace;creating a much healthier planet for our children and grand children."

Hennrietta Clayton was now smiling her infamous smirking but wise smile. She was beaming! "Great idea!" She said. "What will it cost the taxpayers? We all know that's the bottom line, don't we? I think it is a terrific idea. But; can we afford it? What's the estimated cost?"

The General looked at her wistfully. "We can't afford to NOT do it. We have guesstimated an initial $20 million to start. That may prove to be short-sighted. It's an enormous task. Fusion isn't the problem. Sustained containment is the issue. We are talking star-power, with heat of 100 million degrees Celsius or somewhere near 180 million degrees Fahrenheit. Fusion is the way our sun creates it's energy. It 's very hot.

No one so far has found a way to contain the resulting heat that occurs after the Fusion ignition. Physicists can already initiate the Fusion several ways and fairly easily. That isn't the problem. It will take geniuses to figure-out how to contain it.

Fusion is unlike Fission. Fission is dangerous and dirty. Fusion is comparatively safe and clean. There is no danger of run-away melt-downs like Fukushima, in Japan. It takes very little mass to create enormous amounts of energy this way. If we take away the mass or the fuel, the Fusion acceleration ceases immediately. It's very easy to keep it safe and under control.

We have been in touch with the people at Los Alamos already. They are on-board and they already have most of the equipment we need. They would all love to see this Project Starshine take place at those facilities. It will take them back to the glory days of the Manhattan Project. So, we can cut expenses there. The $20 million dollars may be enough. But' it needs to be understood that if we begin with this project we need to see it through to success, no matter what it costs. If more money is needed later, it would be foolish and wasteful to not follow through. There is an outside chance that if we can get just the right genius minds working together, we may bring it in cheaper than that. Imagine the benefits! It's a relatively small investment if we can bring fusion energy to the market place for a mere $20,000,000 million dollars."

The new President didn't need to think long or hard about the proposal. As she sat there smiling, she uttered the question of the day, "**Why in Hell has it taken us so long to do this?** Of course we will get you the money! Why didn't we come up with this idea when Bob was president? It's the idea of the Century. We can not and we shall not fail! I'll be behind it 100%.

I shall appoint committees to oversee the project. I do not want to discover any hanky-panky pilfering of the funds anywhere along the way. Anyone caught misappropriating a dime of the money will be prosecuted for treason! I want that totally understood. $20 million is a lot of money and we will not play favoritism. I will be keeping tabs on every dime and I will personally prosecute any infractions, myself.

So, my answer is yes, General Wilson. Direct the classified forms to my desk on my first day in office. I will approve them as soon as I read them over. Please! No tricks, gentlemen. Are we understood? I love this proposal and the administration will back you 100%. Just keep it honest!"

The meeting adjourned a few minutes later and the President was escorted though the big double doors where her Secret Service detail was waiting to usher her back to the waiting chopper. She felt as though she had emerged from the Lion's Den victoriously. Project Starshine had just become the brightest star on her horizon. It was a great idea and its time had arrived. It crossed her mind that she might grow to appreciate the boys at the DOD, after all. Lieutenant General Frank Wilson was an okay guy!

Chapter 6: A Surprise Visitor . . .

Mr. X had returned from Palm Beach refreshed. He had enjoyed his little hurricane. Monique was still churning as far north as Tennessee but she had been reduced to Tropical Depression status with winds of 25 to 35 MPH. She was no longer a serious threat.

He spent the morning going over the dossiers of contenders who might qualify as the science-team they needed to head-up the project. They each held the correct credentials. He was looking for and hoping for, very special people. The project would require geniuses who could think outside-the-box if necessary. Finding exactly the right people was a monumental and daunting task; like looking for an Oppenheimer or a Feynman in a haystack, or a diamond in the rough.

Sometimes certain people are born into this World with a certain amount of additional ability and drive. Maybe it's their gene pool; or maybe it's the probability laws, or just maybe some superior power is watching, and those people are intentionally guided to help untangle this mess we call earth. Those people become the people our kids read about in the history books a few years later.

They are often the born leaders and they usually do very well in life. For the very special ones, their life-mission will oft times become almost God like. They may tend to operate in the shadows to become the silent protectors of all mankind. Those people tend to rise to the top more easily than other people, simply because they instinctively belong there, as a matter of personality and exemplary intellect.

If those rare individuals are naturally good and moral people, it can often become a good thing for everyone. Thomas Jefferson and Benjamin Franklin were prime examples of that type of personality. Einstein and Feynman were of that type. Mr. X was of that same type and his task was to locate 4 or 5 more of the same, for Project Starshine.

Mr. X had been extremely successful in business as a young man. He had accumulated great wealth and invested wisely. Then as he aged, his money and influence cast him into higher social circles. He began hob-nobbing with varied political figures and power-brokers in Washington D.C. His extreme wealth, intellect and influence had gained him admittance among the movers and shakers of Washington elite. The policy makers all grew to accept him as a part of their crowd. To his own surprise and complete amazement, he soon found that those elite were listening to him and his ideas for the country. He was simply a generous and gregarious soul who wanted to see prosperity for the country he loved. They all admired him for that reason.

It became common for the movers and shakers to look up to him occasionally for the deeper answers. There were times when he would project an idea over dinner or over drinks, only to discover a few days later that the ambitious politician and casual social acquaintance was suddenly ramming his off-the-cuff idea through Congress as a new Bill or law, with their own name on it.

Mr. X was of a different temperament than most people. Rather than feel offended that someone else might be taking credit for his ideas, he saw it as a quiet comfort to be influential enough and in a position to cause good things to happen. It made him feel proud to know that he had been instrumental in creating a little promise and a tiny bit of hope for the people of the country.

Such was the case with the Starshine Project. He had mentioned the idea as a fascinating topic of conversation one night at a pub while drinking with several of the top Pentagon brass.

Within a few days they too had tossed it around among themselves and then officially adopted the idea as a project idea. Gradually, the project became a reality and as it turned out, they unanimously decided to offer him the official management position, purely out of respect.

It was his idea! They liked it and wanted him to direct it. He was trusted and they knew he was a natural organizer. They brought him in (as a civilian) to their private military enclave within the walls of the Pentagon. Almost as a joke, they nicknamed him Mr. X.. They were completely in love with his ingenious idea. The idea had immediately made its way to the Department of Defense drawing boards.

Before he realized it, he had become a fixture and a full-fledged member within the Pentagon crowd of elites. His friends, the powerful Generals within the DOD, went so far as to set him up with a top level security clearance, as though he was a non-identity or mystery-agent, from deep within the bowels of the intelligence community. His true identity remained top secret and his security clearance and badge allowed him to roam freely within the hallowed walls of the Pentagon as one who truly belonged there. Maybe he actually did!

His idea was brilliant, but the same question was on everyone's lips. **Why hadn't the Starshine Project been attempted 30 or 40 years earlier?** It was an idea that should have already been accomplished.

The idea began as a rumor then took on a life of its own. It had created major excitement within the halls of the Pentagon as possibly one of the greatest projects they had ever undertaken.

Mr. X became the privately contracted director for the project and he was offered a six figure yearly paycheck. To his credit, he had completely refused to accept the money! He was a retired billionaire and was quite simply thrilled and completely flattered to be offered the prestigious assignment. He volunteered and insisted that he would take on the job for just $1 dollar a year.

He made a memorable remark as he was offered the assignment:

"Not all things are about money! Sometimes it's about blessings. If a man can't offer a blessing once in a while, he's a mighty poor, poor man."

As Mr. X sat, going over dossiers in his Washington DC mansion, the door bell rang. Mr. X was enjoying a day without housekeepers. He was studying the credentials of the nations top physicists and he was trying to determine which of the ten names might offer the most hope for success with the Starshine Project. He answered the door himself.

As he opened the door he was thrilled to see a lovely lady with hands folded calmly behind her back. She wore an adorable smile on her face. She was wearing a very tight black skirt, about knee length, with heels and a casual blouse that caught the violet of her mesmerizing eyes. Her smile seemed to be powered with its own form of fusion. She was a beautiful late 20-something brunette.

Mr. X felt a little out-classed for a second or two. It was well past noon and he was still in his bathrobe and slippers. He was gracious, though a little embarrassed about his appearance. He was a male and she was beautiful. That alone had him captivated. It would easily gain her his ear and very quick and easy admittance through his door.

"Good morning! To what do I owe the pleasure?" He asked as he invited her in. "Would you like a cup of coffee?" The coffee was already made. All he had to do was pour it.

She answered with an affirmative on the coffee as he followed her in and shut the door. Within seconds a steaming cup of coffee sat in front of both of them. They had adjourned to the couch in the Florida room as she nervously searched for the exact words she wanted to say.

Her gazed wandered out through the open screened-in Florida room and the spacious Olympic sized swimming pool in the background. Beyond that, the estate seemed to go on for acres and acres of manicured lawn and sculpted landscaping. Mr. X was obviously a very wealthy man.

"So, what can I do for you," he asked?

She looked at him shyly then spoke, "sir, I have it on good authority that you may be influential in a particular governmental initiative aimed at tapping the potential powers of fusion energy? I would like to make a proposal. I hope you will find it interesting."

"I'll certainly listen," he said with a smile.

Her smile radiated light almost as bright as the sun itself. "My sources tell me that you may be involved in a directorial sort of way. Would I be correct? My source wasn't certain." She smiled again but this time with a sly question of hope in her lovely eyes. She almost seemed embarrassed to be asking the question. It seemed obvious that she wasn't totally sure of what she was asking or what she was doing there. A negative reply from him at this point would certainly ruin her day.

Mr. X was immediately smitten with her beauty and charm, but he knew he needed to remain guardedly cautious. Project Starshine was supposed to be top-secret! How could she know about it so soon? It was currently just making its way to the drawing boards.

"It *may* be true, but I won't admit to anything like that. Why do you ask?" He found this woman intriguing. She seemed to be about ten levels above him, intellectually. Maybe it was the stylish bifocals that rested pleasingly below the bridge of her tiny upturned nosed. It gave her the look of a totally adorable nerd. Just his type. She appeared to be about 28 years old. He was pushing 46, but she wasn't too young to be out of the question.

He hadn't met a woman like this in many years. He was quietly stirring inside.

She unknowingly melted him with her smile again as she responded. "I suppose I should explain my interest. I'm a recent Ph.D grad from MIT: Experimental Plasma Physics.

For my dissertation supporting my fusion containment thesis I proposed a new idea concerning the implementation of an electrogravitational vortex to help contain and sustain the Helium plasma heat following the initial laser ignition and subsequent fusion reaction.

The design of the gravitational vortex effect will serve to tightly focus the ignition and the acceleration while adding an element of colliding gravitational forces to help keep the plasma more manageable. . . . That's the simple way to describe my idea.

I'm a fan of T. Thomas Brown and I've studied his patents extensively. Truthful, I can't imagine why no one has ever tried to use the amazing technology that he discovered, for the containment of sustained Fusion.

My professors at MIT granted my Ph.D as a result of the feasibility of the idea and the diagrams I presented for my dissertation regarding the concept of employing an electrogravitational vortex housing to help combat and contain the heat problems associated with the containment of fusion acceleration.

They all seem to think my ideas might work, but we need real world experimentation. Those experiments are beyond the capabilities of the MIT labs. We won't be able to pursue the ideas there. Consequentially, these potential game-changing ideas may never see the light of day unless we can gain access to more sophisticated lab facilities.

Fusion and Helium Plasma physics are the future of energy and the World. I've been fascinated with the potentials of controlled fusion since I was a teenager in high school. It seems that I've thought about nothing else, all these years. A check of my academic history and educational qualifications will confirm that. My team and I have also experimented profusely with the Biefield Brown effects with great success. Our somewhat out-of-the-box idea is to enclose a perfectly designed fusion accelerator within a second gravitational container, based on T T Brown's technology. After-all, it's the gravitational extremes within our sun that initiates and contains the fusion, allowing it to burn 10 billion years or more. Our hope is to copy nature on a much smaller scale.

My name is Denise Feywisch and I admit it, I'm a complete nerd." She giggled at that, then continued. "My team and I have specific and definite ideas about how to do the containment. You may find those ideas interesting. It might be expensive and difficult to do our experiments, but my professors at MIT all tended to agree that the idea we have in mind, has definite merit. It might possibly be the ultimate final answer to sustained fusion containment. I'd like to talk to you about it if I may. Would you be interested?"

That's all it took! Now his mind was blown completely. Not only was she beautiful, she was apparently the genius-level intelligent and an out-of-the-box thinker he was looking for. "I think you have my attention completely, " he said it calmly and softly but almost as a joke. On the inside he felt as if she was an angel of God. A Mona Lisa smile commanded his mysteriously intrigued expression. He was more than interested and instantly fascinated with this young lady. "Please continue." he coaxed.

She giggled in a way that was more than cute, though contrary to the discussion. It was an indication that she was a very real and very normal modern girl, but it was not an indication of the very obvious exemplary intellect that quite obviously resided immediately beyond that little-girlish giggle and the mesmerizing violet eyes. This young lady was undoubtedly extremely intelligent.

"Well sir, we did a few limited experiments at MIT that tended to lend credibility to our theories. I worked with and became very close with a team of like-minds. We're all fusion-nerds and we're Thomas Brown electrogravity nerds, as well. We hung out and talked all the time, tossing our ideas around. Anyway, we've have been following the frustrating attempts being made at Lockheed Skunkworks. We completely understand what they are doing, or trying to do. We tend to agree that they seem to have the best, most practical approach with their compact fusion accelerator design .

I mean for example, France has a fusion project in development too and the ITER in France may eventually succeed , but it's so big and expensive. It may not be feasible in a practical way for the rest of the World, or world-wide implementation for 30 or 40 more years. That's just too long to wait!

We need this technology right now! Their time-line puts them completely behind the eight-ball. We need this technology immediately . . . or we may not be here in 30 years! We simply must discover something other than fossil fuels.

The Skunkworks project just makes more practical sense and if it succeeds it will be incredibly awesome. It is small enough to be carried to where ever it is needed and it's small enough to facilitate easy and economical experimentation.   That's the thing about the Skunkworks idea! It's the least expensive and most practical. They are on the right path. I understand that they have been dealing with setbacks caused by the heating problems of containment.

Those problems seem to be the core issues that have plagued and thwarted fusion attempts all along. My team and I need a place and opportunity to experiment with our electrogravitational vortex idea. It might just work, if given a chance. That's why I'm here. I think you may have the influence we need to help us try our ideas for real. We hope to offer the United States a fighting chance in the struggle for clean and economical energy."

She was on a roll. "You see sir, if an electromagnetic induced gravitational vortex could be initiated at the onset prior to and before the actual firing of the ignition lasers, the idea would be to insulate, then catapult the deuterium and tritium atoms to coalesce into a sharpened plasma focal co-existence within the . . .

Mr. X held up a hand to silence her momentarily . . . He was grinning, and laughed openly at his own ignorance. "Whoa! You're already a mile or two over my head. Needless to say, I can immediately see that you know exactly what you are talking about. The problem here is the fact that, I'm totally clueless. I'm adequately impressed though. You have made your mark on me, though all I can do is get you in the door. You will be on your own from there.

I will want to take you to speak with the people who will know and understand exactly what you are talking about. If you can make the same impression on them that you have just made on me, I think you may have a bright future on this top-secret team.

Would you be available to chat informally with a few of the higher-ups in the project, today, . . . like, . . . later this afternoon?"

"I certainly would be." She giggled the adorable little-girl giggle again. It was absolutely infectious and endearing. "I was hoping you might say something like that. Name a time and place. I'll be there and I'll be very well prepared. They will not turn me down if they'll only give me the floor for 15 minutes. I think my team and I are definitely onto a concept that will positively work. It begs to be tried. History may never forgive us if we don't try."

Mr. X reached for his cellphone. "Tell me your name again. We'll get a couple of people on the horn and see what we can setup. I believe we can do it in a 100% informal way; at least for today's initial meeting. I'll get a few of the key individuals to join us for a few drinks about Happy Hour. I'll introduce you to them and give you my two thumbs up. You can take it from there. I have a feeling you'll have them eating out of your palm in 10 minutes or less.

"My name is Denise Feywisch" she said.

Mr. X hit the speed-dial on his cellphone. He was one of the few people in the world with with all the top brass at the DOD, on speed-dial. They were his closest personal friends.

"Frank! What's up for happy hour today, old buddy? I've someone you will definitely want to meet!" He shot a friendly quiet smile to Denise as he listened into the cell. "Okay! 5:30 at the Commanders Room. Can your bring professor Enderman with you? Tell that cheap old skinflint I'm buying; that will get him there for sure." He chuckled. "I believe we will need his credentials and his brain for this little gathering. You and I may not understand much of the conversation at all. It involves what sounds to me like a very interesting proposal for the . . . Project." He was silent for a few seconds as he listened. "Great! We'll see you both at 5:30, Frank. Bye!"

He smiled at Denise. "I hope you are free at 5:30. We'll ride over in my old Stutz. That was Lieutenant General Frank Wilson at the DOD. He's a three-star General in charge of research and development. In other words, he's the top boss of a new top secret fusion containment project. He will be bringing Dr, Enderman, our plasma physics genius and fusion energy consultant, with him. Dr. Enderman will understand the feasibility of what you are talking about, where Frank and I might not.

Dr. Enderman is a sweet and lovable old guy. He's the classic image of an absent-minded genius. Sometimes he forgets to tie his shoes, but he can recite pi to 300 digits. We all love to tease and kid with him but he's an absolute mental giant with a heart of pure gold.

This project is extremely top secret by the way. How on earth did you hear about it, and what led you to me?

She looked embarrassed and at a loss for words for a few seconds before she shrugged her shoulders and spoke, "well, I sorta have a friend over at the DOD who is on the inside. He threw your name out as a wild guess. Are you actually the man so many refer to as Mr. X?"

"Yes." He responded. "That label has stuck. The guys at the Pentagon hung that dumb moniker on me, since I am on the inside, in a best friends sort of way. They all know and like me. Since I don't wear brass on my shoulders and I'm not really elected or a part of the DOD, they nicknamed me Mr. X and then created a classified Army Intel identity and security clearance for me so that I can come and go at the Pentagon, without a lot of hassle. They have set me up as though I'm some mysterious no-name Mr. X, from deep within the Army intelligence community. My real name is to be kept top secret, primarily because I'm actually a nobody! We don't want anyone else to know that!" He laughed at how preposterous it all sounded. The fake moniker and security clearance gives me cred that I wouldn't ever have. If the admin people actually knew I'm just an average Joe with good friends on the inside, they would probably boot me out the door. Everyone calls me Jim when I'm talking one-on-one. It's also a fake name, but Mr. X raises too many eyebrows. Please, just call me Jim too. We do need to keep my real identity a secret. Sometimes I go by Jim Axe."

She smiled knowingly, and didn't press the issue, "Okay Jim. It'll be Jim from now on. That's all I need to know. I really don't need to know your real name. No need to elaborate. It's funny, but all I had when I came over here today was your suspected address. My friend, or source, was guessing 100%. He didn't know for sure if Mr. X would answer the door or not and he didn't know if it would lead to anything. So, I was simply grabbing at a lonesome straw when we first said hello. I had no idea of what I was going to say. I was just winging-it and hoping for the best. It looks like we got lucky and rang the bell!"

Jim X chuckled as he finished off the coffee. "Look," he said "I need to get out of this bathrobe and get into something presentable. How about if I flip on the TV and leave you here for a few minutes while I freshen up and get into some street clothes."

"Sounds cool to me," she said, "do I need to know anything special to work the Remotes?"

He flipped on the TV, then handed her the only remote she would need, while showing her how to reach the Channel Guide. She had it figured-out in an instant. He then disappeared upstairs to get ready for the meeting.

As he milled about before the shower, he pulled up the upstairs computer terminal and accessed the dossiers that Captain Dugger had sent. Sure enough, Dr. Denise Feywisch Ph.D was number 8 on the list. It mentioned that she worked with an impressive team that included three other young Ph.Ds. All four had captivated their professors with their dissertations and their theses concerning theories of electromagnetic gravitational vortex containment within compact and portable fusion reactors. It sounded as if they might be 100% perfect for the project. The fact that her name was on the dossier list had just confirmed his vote.

Chapter 7: Happy Hour with the Brass.

At 4:45 Pm, Jim and Denise walked out through the kitchen door into the spacious 4 car garage at the back of the mansion. Jim hit a button and one of the four automatic garage doors lifted open. He had decided to take the Stutz Bearcat out for a run on this special day. He drove it only rarely, but it seemed to be a fun idea and just maybe he was trying to impress the little lady. The Stutz DV32 was a beautiful old classic relic from 1932. It had a canvas top and all the elaborate trimmings of a luxury sports car. It was a precious antique.

Denise lit up like a Christmas tree. To ride in such a car would be a once in a lifetime experience and she knew it, even though she wasn't an antique sports car aficionado.

Jim smiled proudly as he spoke. "The Commander's Club is only a couple of miles from here. This old buggy needs a little work-out. I haven't taken her out in a while. Hop in, buckle up, shut-up and hold on!" He chuckled as he recited his favorite line, then cranked the engine. It turned over once then fired, popped and roared powerfully to life.

Denise was forced to show quite a lot of leg as she climbed in and buckled the seat belt. Jim took note approvingly with a gregarious, happy smile. Their eyes met, then dilated with instant mutual admiration. He noticed but she didn't.

Within seconds they had wheeled out the driveway and hit the hi-way. Everyone on the road felt a need to wave admiringly as they passed the old sports car. It was bright yellow and difficult to ignore. It made Denise feel like a celebrity. She hoped she wasn't showing too much availability, but she was definitely impressed with this guy known only as the mysterious Jim X. He seemed like a perfectly wonderful guy.

Mr. X proudly escorted the lovely young female scientist into the Commander's Club. Several of the Pentagon Generals were already nursing a first happy-hour drink. They looked-up and smiled with raised eyebrows and mini-pancake eyes. General Wilson and Dr. Enderman hadn't yet arrived. Jim X introduced each of the Generals to Denise and explained why she was there and with him. Each of the Generals appeared to be genuinely impressed based entirely on her appearance.

A few minutes later General Wilson appeared with Dr. Enderman at his side. Mr. X introduced Denise to them as well.

It only took a few minutes before Denise and Dr. Enderman were engaged in a deep discussion concerning the Skunkworks project and her ideas for the inclusion of an electrogravitic assembly shell for sustained containment of the fusing process. She went immediately into an explanation of how an electromagnetic gravitational shell surrounding the Toroidal container assembly would provide extra and very powerful containment by inducing a 10 G gravity effect in addition to the helical magnetic fields within the Torus . It would assist in providing an aperture focusing effect of the unruly plasma. It was a simple leap of logic to see that the gravitational assist might be the ticket to help with initiation, containment and perpetuation of the fusion process, similar to the way the Sun accomplishes it.

The objective of the electrogravitics was to manipulate powerful G-forces to both control the plasma within the Torus while economizing on the necessary power for the initiation process.

The idea would be to use a gravity assist to continually drive the fusion plasma into sharper focus and into itself, while facilitating and sustaining the internal temperatures. The hope was to achieve and maintain the extreme levels needed for fusing, while at the same time initiating an insulating and cooling effect, externally. The additional inward and strong gravitational forces would also help induce the necessary pressures directed at the substantial coils, then provide an extra needed push to bring and hold the plasma within the helical magnetic field. This would facilitate induced ion collisions and increased odds at allowing and causing fusion to occur, then focus and contain the fusing plasma for longer and more productive periods.

Dr. Enderman was analyzing her ideas scientifically and looking for errors in her thinking. Her electrogravitic ideas set his own imagination ablaze. It was something no one else had tried as far as he was aware. It seemed to him to be a plausible idea.

Before long she had convinced him that the only true way to know if the idea would work, would be to try it. It deserved experimentation and that was exactly why she was there. She knew the MIT team could do it and she was hoping to be chosen for the Project Starshine team.

He seemed to be accepting her out-of-the-box thinking, at least until it had been proven right or wrong. He couldn't conjecture any reasonable reasons why it wouldn't work. Her idea was deserving of experimentation. He knew the Skunkworks team had experimented with T. Thomas Brown's electrogravity when the Stealth B2 Bomber program was in development.

All of that was considered very black and extremely classified. It was something that was only talked about in whispers, even among themselves. He only knew about it because he was on the inside at the DOD. He had overheard the whispers and had seen interesting pieces of documentation that referred to it optimistically as if it had worked successfully.

Denise explained that she and her team had developed a simplified stair-stepped series of testing experiments that would prove or disprove their theories in a practical and inexpensive way. They had already experimented with the electrogravitational ideas separately and with great results. She already knew they could create and direct a focused, concentrated inward sphere of 10 G gravitation to assist with plasma control within the containment vessel.

That was another sell-point in her favor. She was well prepared as she sketched a diagram on a bar napkin. It helped to illustrate the functions of an outer containment electrogravitational housing assembly. The elaborate sketch explained in detail how it functioned to create the 10 G gravitational effect.

Denise went on to explain that she had already gathered a team of like-minds while at MIT. She hoped that if she was to be chosen for the project, she would like very much to bring the entire team with her. There were three other people on the MIT team, all with Ph.Ds to their credit. The complete team consisted of one other woman and two men.

It wasn't long before Dr. Enderman was defending her ideas in a grandfatherly way and in hushed tones, to the various Generals and Pentagon elite at the table. He obviously loved Denise immediately and he was blown away with the potential that existed within the gravitational idea. It appeared to suggest the missing link that had been lacking in all the failed attempts at sustained fusion in the past.

It was apparent that Denise was just as enamored with Dr. Enderman. He was an endearing soul.

Thirty minutes after that, it appeared that Denise had been completely accepted by everyone there. If the Doc said she was in, then she was probably in! It meant that she and her team would most likely be accepted as major part of Project Starshine. Of course, the other members of her team would all need to be scrutinized by Dr. Enderman as well. If they too could impress him as well as Denise had, they would all need to endure Class A security checks and they would then become accepted as the project's scientific team.

All of this possibly meant that Denise's life had just turned on a dime for the better and as a direct result of meeting Mr. Jim X, earlier in the day. She was thrilled and exuberant at the prospects.

She and Mr. X had a total of three drinks each. Before long she whispered in his ear that she really needed to be going. She knew her drinking limits and she didn't want to botch everything by drinking too much.

The couple casually excused themselves from the group with and exchange of phone numbers around the table. She was invited to bring her entire team with her on Thursday for a meeting at the Pentagon with Dr. Enderman and the other staff members of the project. If the remainder of her team were acceptable, they would all undergo security checks. The entire team would then be sent to Los Alamos to begin work on the Starshine Project as soon as possible.

As Jim X and Denise were leaving the Commander's Room, Jim mentioned that he was hungry. He had planned to eat at the Commander's room. Denise was so excited about what had just transpired, she offered to buy dinner. "Do you like Prime Rib" she asked? "I know a great place and I'm buying. Are you game?"

Jim felt edgy about allowing her to buy. He was worth a fortune and he knew she could sense it. It was a great gesture on her part. "Sure! Let's do it. Where is this place?" He asked, as they climbed into the Stutz and fired the 8-cylinder engine. It roared to life with a pop, then idled smoothly.

"It's called Martinos. Take a right out of the parking lot then another right at the next intersection and you'll see it over on the right. I go there a lot. They serve a 16 oz Prime Rib for $15.00. It's to die for and pretty hard to beat. I would just like to reward you for helping me this way. Getting to know you may have just created a major upturn in my life, and future. I can't thank you enough." She was bubbly and totally lovable as she spoke.

"I apologize for the quick exit. I wanted to get out of there rather than take a chance that I might drink too much and ruin a my good first impressions. I know me better than anyone and I have a bad track record with alcohol. I don't always handle my booze as well as I would like." She blushed and held both hands over her eyes, giggling and shaking her head in embarrassment.

Jim X was smitten and Denise seemed to be available. He was seeing some very real possibilities with this lady. He would have happily followed her to the moon and paid the tab as well. There wasn't any way on earth that he could refuse her offer.

Soon they were in Martino's enjoying delicious prime rib and more causal drinks. Before they realized it, they had each imbibed three more drinks. She was drinking Frozen Margaritas and he has having his usual JB & Coke. Jim noticed that she was becoming more than a little tipsy before they decided to head back to his mansion. He was feeling the booze as well. He knew to get the Stutz home and safely in the garage before drinking any more.

As they returned to the mansion Jim mentioned that he had a full jug of Jose Cuervo at the bar at home and all the fixin's for more margaritas if she would like another drink or two. To his surprise, she accepted and it became a full-on party of two at a poolside patio table. They ended up nude in the pool a couple of hours later. She had become much too inebriated to drive by that time.

The two became romantic and she stayed all night. They were actually a match made in heaven. She was a 100% normal female in every way, except she was sporting an IQ that was well above average. Both became equally enamored, each with the other. Cupid had simultaneously landed his arrows on both targets.

Chapter 8: The Starshine Team Emerges.

By Thursday, Denise was staying at the mansion. Jim X was pleased beyond words. She felt the same way. There weren't any games to be played. The two had connected like super magnets with an electrogravitational assist. They appeared to be meant for each other.

She contacted all her team mates and told them the big news, that they all might be chosen to go to Los Alamos to begin work for the Pentagon. They would be working on their mutual ultimate dream, the fusion solution.

They all agreed to meet at the Pentagon on Thursday morning at 9:00 AM for a special assessment and evaluation meeting with Dr. Enderman, General Wilson and the other staff officials. If they were accepted into the project, they would soon be on their way to Los Alamos in New Mexico permanently, or at least until the completion of the project. Each member of the team immediately agreed to being excitedly available to take the job if they were accepted. It was their dream come true.

On Thursday morning, Jim and Denise took the brand new 2016 Mustang over for the meeting. Jim wanted to drive it. It had arrived just a few days earlier and he hadn't even sat inside it yet.

Jim also owned a new Lexus and the 2015 Cadillac Escalade Limo in West palm. He had the Chauffeur on a weekly retainer. The Chauffeur's job was to travel back and forth keeping all the automobiles in tip top shape. He occasionally drove for Mr. X, when requested.

But Jim was totally capable of driving for himself, and he enjoyed it. The Mustang was an absolute work of art. So he allowed Denise to make the choice as he stood humorously patting the beautiful Mustang with one eyebrow raised. She giggled, caught-on quickly and they unanimously decided to take the Mustang for the short ride over to the Pentagon.

As they arrived at the Pentagon parking lot, Denise met with her friends at a prearranged parking spot and then introduced her teammates to Jim . . .

There was Audrey Reynolds a striking 31 year old Plasma Physics Ph.D. She had flaming red hair and was built like a concrete out-house. She was arguably as beautiful as Denise.

Jim was shocked a little as they were introduced. He hadn't expected such beauty and intellect to be coupled together in just one person.

There were two guys on the team. Gus Jenkins was an All American 'Jock' type guy with a toothy Tom Cruise smile and buff physique. He also held a Ph.D in Experimental Physics at age 29. Denise claimed he was a regular MacGyver; the type of guy who could build almost anything with just a rubber band, a toothpick and a paper clip.

Finally there was David Zimmerman, approximately 30 years old. He was a genuinely handsome type, in a nerdy way. He wore glasses on a thin, almost skinny frame. He wasn't exactly buff, but he didn't look unhealthy either. He also held a Ph.D in Experimental Physics. Denise introduced him as MacGyver #2. This team was the cream of the crop as far as intellect was concerned. They had all finished at the top of their class and they were best friends too. They made a very impressive team.

As Jim spoke briefly with them before the meeting, he became immediately convinced that Project Starshine had just found the people who could and would push the ambitious project over the top. Those four names were likely to become four of the most famous names in fusion physics. They could expect to reach star status as physicists, if they were to succeed.

Within minutes the group was ushered into the DOD office deep within the Pentagon by two snappy looking Marine guards. All four were introduced to the top level Generals and Dr. Enderman.

As a group, they were grilled, drilled, diced, sliced and asked a passel of personal and sensitive questions. The meeting went very well. Before it was over, the entire group had been accepted and they were signed as the primary team committed to the development of fusion energy with the Starshine Project. They all began the processing for Class A security clearances right then. The processing would take several weeks before their security clearances would be complete.

Jim X volunteered to personally fly them to Los Alamos aboard his private Learjet 45XR once they had all obtained their security clearances.

The time passed quickly as they waited. It meant big changes in all their lives. It was essentially an indefinitely permanent move to New Mexico. The following weeks became a matter of closing accounts, dropping classes, deciding on what to take with them and what to leave behind. Each would be allowed 200 lbs. of luggage aboard the Learjet. It would mean they couldn't take everything they owned.

Both of the guys decided to drive to New Mexico. They wanted to have their own cars to get around in, once they were there.

The girls weren't very concerned about that yet. They figured they could bum rides with the guys if necessary or they might return later and bring their own cars down to New Mexico if it turned out that they needed them. They wanted to fly on the jet.

By the time the team had all acquired their security clearances, Jim and Denise had become virtually inseparable.

Money was never an issue for Jim. He decided to buy a home in Los Alamos on arrival. He needed to be there consistently as the director. He would find a suitable place, close to the labs and stay for the duration. Denise would stay with him. The new adventure was shaping-up nicely.

They were all anxiously awaiting final word on their security clearances. Finally, one day 7-weeks later, they were successfully cleared and ready to make the move.

Hennrietta Clayton was officially sworn-in on Tuesday, January 4th, 2017. Her very first act as the new President was to quietly approve $20 million in black budget funding for 'Project Starlight'.

Chapter 9: Sweet Home Los Alamos.

At 2:00 pm on Wednesday the Learjet landed at Los Alamos Municipal Airport.   Jim had sent the Chauffeur ahead in the Mustang. He was waiting in the reception area. The Chauffeur helped them unload their luggage from the Learjet into the Mustang. Then, Jim sent him back to DC on the Learjet, and paid him in advance for a two month vacation. Jim wanted to keep the chauffeur on a retainer but he wouldn't be needed in Las Alamos for at least two months.

As soon as they all bid farewell to the Chauffeur, Jim and the two girls set out in Los Alamos, looking for a place they could call home while they were there. Within three hours they had found a nice 5-bedroom 3-bath home with a pool at 3835 Arizona avenue. Jim paid cash for it with a check. It was only $950,000. He laughed about the price, saying he would probably make money on it when he decided to sell. It was a bargain bank-repo that had sold in 2006 for nearly 2 million. The luxurious home was a beautiful place on a hill with a decent view of Las Alamos. Caballo mountain was just out their back door.

The Las Alamos area was a lot more arid than any of them had expected. After all; it was in the middle of what was essentially a desert. It was very dry and mountainous but beautiful in its own way.

It would be a couple days before the team could begin work It and would take that long for Gus and David to arrive by car. So, the trio consisting of Jim, Denise and Audry decided to explore the small town. They found a good restaurant not far away and stopped to eat.   Afterwards, the three drove around for quite awhile getting accustomed to the layout of the town. It was tiny when compared to DC. The new house was about 2 miles from the lab where they would do most of their day to day work.

The eager group stopped by the lab and became acquainted with a few of the people who worked there permanently. It was good to discover that a small army of dedicated and ambitious helpers would be working with them to do the heavy lifting, run errands and whatever. The boys at the Pentagon had taken care of everything.

Everyone at the lab had been expecting them and they were already preparing the setup for the new fusion experiments. A good portion of initial work had already been completed before they arrived. Denise was thrilled to discover that she and the team could begin working on the electrogravitic housing assembly right away. The regular crew had already built a basic Fusion reactor Stellerator very similar to the Skunkworks project. They were an exceedingly efficient bunch.

Gus and David both arrived at the same time two days later. They had traveled together on the road all the way. It appeared that they both had enjoyed a great, fun trip. They laughed about the fact that they had stopped over in St. Louis and got smashed. The hangovers had caused them to get a late start on their second day.

After a discussion with Jim, they too decided to stay with him and the girls at the new house on Arizona avenue, since there was plenty of room. The group was just now beginning to get a feel for how much Jim was actually worth. They would all live rent-free at Jim's place. He would be picking up the tab with a smile. Money was simply not an issue for him. His investments were making him ten times more money per day than he spent. He never worried about money.

Everyone on the team had signed on at $120,000 per year. So, as a group they were all sitting pretty. All they had to worry about was the not so simple task of containing and sustaining a fusion reaction. Converting the energy to electricity would be easy. That technology already existed. The reactor would simply heat water, create steam and spin turbines to create electricity.

Dr. Enderman flew in a day later. He would stay and take charge of the project. The Dr. was considered the chief adviser to the team. He was a wonderful old guy, well into his 60s. He was brilliant too. Something about him reminded everyone of Albert Einstein. He was forgetful at times and he mumbled equations a lot. In other words he tended to think out loud and he was usually doing calculus in his head as he did other things at the same time. He was a man to be revered and loved. His hair was usually unruly and unkempt and he had a passion for smoking his hand-carved ivory pipe. Some said he was known to load it with marijuana from time to time but he wouldn't admit to that. He had grown up during the Hip-happy days of the 1960s and some of that culture was still a part of him.

Jim still had an empty 5th bedroom at the house, so he invited Dr Enderman to stay with them too. The good Dr. accepted for the time being but thought he might prefer to live alone in the government housing. The government housing was very nice and within walking distance of the lab. So, as it ended up, he would visit with the team from time to time but he actually preferred living alone. It was just his nature. He had lived alone since his wife had died five years earlier. He claimed to enjoy the solitude, saying it helped him keep his imagination sharp. He did stay in the 5th bedroom for the first two nights, but by the 3rd night he had snagged a very nice 1 bedroom apartment at base housing.   It was provided free as a small perk to help convince him to stay with the project until its completion. The apartment was just two blocks from the lab and the Pentagon would pay his rent.

Chapter 10: The Real Work Begins.

Bright and early the next Monday morning the team showed up to begin work at the lab. The first two hours were spent meeting all the regular staff, chewing the fat, learning the layout of the lab and getting organized, before they could begin.

Denise had piles of folders and flashdisks loaded with diagrams and equations that would help lead them to the final design of the electromagnetic vortex assembly. She devoted half the morning to entering all the data into the lab's super computer terminal.

Another great idea was occurring in her mind as she sat organizing her work space. She realized that the laser vortex ignition would reduce the size of even the Skunkworks model. It might potentially mean that the compact Fusion reaction arrangement would be small enough and simple enough that two or more plasma containers might work in sequence. One would do work while the other(s) cooled, then the sequence could switch back or possibly go to the third and possibly a fourth container. The idea might help create a continual and sustained Fusion reaction in the focused vortex, 24/7/365. Enormous amounts of energy would be possible that way. The idea seemed crazy at first glance but her deeper intuitions told her it was worthy of more thought. There might be a way to keep the delicate balances and make it work. The small epiphany seemed above and beyond anything she had thought of before.

But, it was much too soon for that. She made a note on her computer to try it later. First they would need to implement the gravitational housing designs. She had several really good vortex design ideas, and the other team members were working on their own. All ideas would eventually have their moment in the sun until a winner could be decided upon.

By mid afternoon the group came together for a meeting. They examined the partial reactor that had already been setup and built by the resident employees. It was incomplete but patterned almost too much after the Lockhead project. They would need to make changes, but it was a good start.

The team always worked together. They rarely argued, but they had a tendency to tear each others ideas to shreds before voting on the ideas that survived.

Dr. Enderman's unique and lovable personality became invaluable. He listened to each idea, nodding encouragement to each individual, then he would cast the deciding vote anytime the group was deadlocked. It didn't take him long to prove his value. He always backed his vote with sound reasoning, then invited more rebuttal. His final vote usually ended the argument.

Eventually a workable plan would emerge that seemed doable. They would put that idea on a to-do list, then move on to the next idea, the same way. After a number of potentially good ideas had begun to fill the list, they then debated until one idea rose above the others, to become their first experimental attempt.

By six o'clock, they had settled on what everyone hoped might be the best of all their combined best ideas.

Denise's initial idea had always been to funnel electrogravitational and magnetic forces into a central pinpoint of very sharp focus then stabilize as ignition was initiated with lasers. Injectors would propel the tritium and deuterium in precise amounts with precise timing in such a way that the plasma would form but remain pinpoint focused inward on itself and completely away from the housing assembly as magnetic and 10 G repulsion effects continually streamed the plasma to a sharpened point much like a tornado; keeping well away from the rest of the assembly. This would result in duel insulation and protection from the plasma heating. It went a lot deeper than that but they all knew exactly what needed to happen. As a team, they focused and they worked. Each assumed a task automatically.

It took almost a month to build the exact design they wanted to achieve. There were problems. Gus and David tackled each one like competing MacGyvers. When one problem was solved it often created another. Gus and David would tackle it together, with the girls pitching in every way they could. Gradually 'their baby' began to take shape. It was beginning to look like a hopeful prototype.

Before they would jeopardize the entire assembly in a go-for-broke shot-in-the-dark, they needed to do a number of important tests.

Tests and more tests of every sort were necessary to be certain of the integrity of virtually every part of the new assembly. Some of the tests couldn't be accomplished in the Los Alamos labs. They had a list of five separate tests that could only be done safely at the Sandia labs in Albuquerque. It meant that they would all need to go there.

Exactly six weeks weeks after they had begun, the Sandia testing facilities had been granted. Their initial vortex assembly tests could be done over a specific 3-day period. It seemed to be a rushed schedule but they thought they could pull it off, though it might mean a couple of 20-hour days. They were okay with that. They tended to do 20-hour days all the time. When things were moving nicely, sometimes their creative minds just didn't want to disengage. It seemed that everyone on the team was the same way in respect to that. Life is like that when you love your work.

Sandia labs was the next stop in their journey.

Chapter 11: Spies!

They drove two cars to Sandia. Gus, Denise and Jim took the Mustang. Dave, Audrey and Dr. Enderman went in Dave's Subaru. The other employees from the Los Alamos lab would carry all the necessary equipment down to Albuquerque in a military transport truck.

They all traveled together as a happy little convoy and the girls talked on cellphones most of the way. Jim had to remind them twice that the project was top secret. They were not to mention any specifics on the phones. If they mentioned anything about the project at all, it needed to be in coded wordings. He had heard Denise say Project Star . . . two different times before she caught herself, smiled at him and changed the wording.

The prime topic they talked about was the fact that the Sandia Laboratory was a division of the Lockheed Martin Corporation. Skunkworks was also a division of Lockheed. In other words, they were on the home turf of their quasi-competition. Never mind that their reactor design was based loosely on the Skunkworks design. It was a potential can of worms. They would need to be very careful about talking or showing too much to the wrong people.

Albuquerque was under 2-hours away. It was a short trip.

A short while later Denise called her brother Eddy who worked as a computer security analyst for the DOD. Eddy was in a sense her secret source on the inside. He was in the know about everything.

"Hi Sissy," he responded on the second ring. "I've been meaning to call you. First I want to congratulate you and your friends. I'm so proud of all of you. How's Los Alamos?"

Denise and her brother were close. They were just a year and three months apart in age. He had always been her protector when they were growing up. "Well it's really dry out here but comfortable most of the time. We love it. Everything is going well. We have to go to Albuquerque to do some tests. We'll be there three or four days. How is everything there?"

"I'm just fine and I already knew you were on your way to Albuquerque. I have my ear to the ground all the time. Everything goes through these computer systems. I picked up a little piece of news that may interest you guys. I have to be careful what I say on the phone. Let me say that you are being watched, by people other than the DOD. There are people, big people who are worried that you guys might actually succeed at what you are trying to do. Keep your eyes open and watch your back, there in Albuquerque. Watch for spies. Do you remember the way we used to communicate when we were kids, Denise? The cipher? I will be sending you an e-mail with specific details I don't want to say on the phone. You will need to figure it out when you get it. It may be important. Sometimes I worry for you. What you are doing could now be very dangerous. There are some really big people who may want you to fail. Do you know what I mean?"

"Yes I do. It has been pretty calm so far. I will keep my eyes open and I do remember the cipher. I'll be able to read the e-mail. You keep your eyes open too. Notify me of anything I should know. Bless you. You are a great brother. I should get off of here, we're coming into Albuquerque now. I love you. Stay in touch." After he said goodbye, they disconnected.

On arrival at Kirtland Air Base, the team went immediately to the Sandia labs and found the specific testing labs they would be using. They were directed to the TA-IV section. The crew began setting up the needed equipment as Denise and Gus supervised. Everyone wanted to get a head start since their allotted test-time was just 3-days. They would show up early the following morning to begin the first day of testing. Their plan was to test until they physically dropped from exhaustion.

As they prepared the equipment, several of the Sandia employees loitered about, asking far too many technical questions. They were just a little too interested in what the team was doing. It made the entire team nervous.

Dr. Enderman was especially concerned and upset. He put in a call to the Pentagon to confer with General Wilson concerning the precarious lack of security in this unique situation. Then he cocked an eyebrow at Jim who was officially the director of the project.

It fell on Jim's shoulders to set the two over-eager employees straight. He realized it immediately. He tried to handle the situation diplomatically. The two men were informed in no uncertain terms that the project was classified top-secret military.

Their rebuttal to Jim was the simple fact that it was their job to see to it that all safety precautions would be followed during the upcoming tests. They would be there the entire time as the team did its testing. That piece of news was very unsettling for Jim and the entire crew. He escorted both men, with a friendly but threatening smile from the lab facilities anyway. They were completely miffed at the treatment and argued with him briefly before he very kindly showed them the door. They considered the lab facilities as their territory. The Starshine team were the interlopers! A smiling Jim escorted them from the building anyway. They left in a huff. It might mean that he would need to call in the big guns from the Pentagon. Something just wasn't right about the whole setup. Security was much too lax.

Within two hours the team and crew had everything setup and ready to go for the next morning. They had made reservations at the nearby Extended Stay America Hotel a few miles away. Jim had paid for 5 rooms for the entire week. They all headed for the Hotel and some precious rest. It had been a long day.

Just as they reached the Hotel, Denise became totally distraught. She had misplaced her purse. It contained flashdisks that held extremely sensitive diagrams for the electrogravitic assembly of the reactor. She was in a complete panic.

Jim calmly wheeled the Mustang around and headed back to the labs. Surely Denise had left her purse in the laboratory somewhere. It wasn't like her to do things like that, but it had happened. Soon they were back at the lab unannounced.

As they entered the lab facility where their equipment was setup they were met with a big surprise. The two men that Jim had escorted from the building were there. All the lights were on and focused at the partially assembled reactor. They both had their cellphones out and were snapping pictures as Jim and Denise walked in unexpectedly. This was bad! The two men were obviously spies!

Jim was usually a very calm man by nature. This situation rattled him and Denise to the bone. Their top secret project was being exposed in minute detail to whoever those two men worked for.

Jim was cool. He held up one hand and speed-dialed General Frank Wilson's private cell number in DC. It rang three times before Frank spoke, "Jimmy my boy! What's up? I was just getting ready to hit the sack. What can I do for you?"

Jim explained the situation. The General was livid. "You say they are both military? I want to speak we each of those men personally. But; hold for a second!" He put Jim on hold for about two minutes. Jim had no idea what was happening. Then after a few minutes he came back on the line. "Put them on your cellphone, right now . . . one at a time. I want to have a few words with each one of them."

Jim handed his cell over to the largest of the two men. The guy was a 220 lb monster of a man in his mid-30s. He looked like the sort who might thoroughly enjoy a good John Wayne barroom brawl.

The man spoke into Jim's cell very bravely at first. "Hi! I am First Lieutenant Gerarld Adamson. How can I help you?" He said it with all the authority of a General.

It was almost comical to watch his face begin contorting. At first, he was in total disbelief. Then he became immediately humble, then even more humble, "I'm so very sorry sir. I do apologize. Sincerely! Yes sir, I do understand completely. Yes sir, we will do exactly that. Again, we apologize. Court Martial? Sir, it is simply a misunderstanding. We will rectify it. We had no idea. Yes, sir. Okay sir. I will do that sir."

He then handed Jim's cell phone to his partner and silently made a motion as if he was slicing his own throat. His eyes were the size of hub caps. They were both in very deep trouble and he knew it.

The scenario played out again as the General reamed the second guy in the exact same way. Their conversation lasted a good two minutes. Meanwhile, without a word, both men voluntarily and very quietly handed over their own cellphones to Jim, with sheepish looks on their faces.

Just as the second soldier clicked the off button and prepared to hand Jim's phone back, a squad of four Air Police officers came through the door with guns drawn. Apparently General Wilson had called them during the moment while Jim was on 'hold'.

One yelled, "hands in the air! All of you! Do not move! Weapons were drawn and they meant business. Apparently the General had called them directly from DC.

Everyone in the room raised their hands. It took a minute for the APs to sort through the situation and separate the good guys from the bad.

Jim, being of a calm and controlled nature even tried to step in and vouch for the two men, a little. Both men had immediately surrendered their cell phones to Jim and apologized sincerely. It meant that Jim had all the pictures they had taken. As he saw it, no real harm had been done.

Soon the two men were in handcuffs and they were being escorted to the Stockade. The Air Police were apologetic to everyone there, as they explained that the incarceration order had come from the head of the DOD himself. If Lieutenant General Frank Wilson wanted these two men in the Stockade, that's where they would be until further notice. The Air Police were under strict orders from above and they had to do their duty. The two men would be sleeping in the Stockade until further notice. Case closed.

Chapter 12: Wrath of the DOD

Denise found her purse exactly where she had left it. It did not appear that it had been tampered with. She exhaled a sigh of relief.

She and Jim were a bit shaken after all the excitement, so on the way back they stopped at a 7-11 and bought a 12 pack of Coors Light before returning once again to the Extended Stay America Hotel. After reaching the room they settled on the balcony of the Hotel for a refreshing beer before bed.

As they sat on the balcony, Denise fired up her laptop and checked her email. There was a message from her brother Eddy. As he had warned, it was coded in an old cipher they had agreed on when they were kids. It was a simple displacement cipher, but on the computer, they used the entire ASCII code. It made the coded message much harder to break. Even if you knew the key, it was hard to decipher. Without the key it was almost impossible. They had always used the same key. She broke the code and deciphered the message after struggling with it for a little while.

In essence, the message stated that the kingpins of the big oil companies in Houston were in cahoots with some of the people at the DOD. Eddy had seen e-mails detailing the Starshine Project and essentially someone unknown from within the DOD was keeping the Houston Kingpins well informed about the progress of Project Starshine. Eddy claimed he already knew they were on their way to Albuquerque because he had seen it in the communiques from the DOD to the people in Houston and someone over at the NSA. Houston's reply to that, had been that Houston was planting people at Sandia who would be watching.

Denise read the deciphered e-mail to Jim. They were both deeply concerned, but for Eddy's safety they couldn't confer directly with the General about it. Eddy would be in serious trouble. He was breaking some serious rules to gather the information. They would need to remain quiet about the new knowledge.

From where they sat on the balcony, they could easily see the Mustang parked in the parking lot below. They sipped the beer and talked mainly about the project. Denise was trying to explain her admittedly out-of-the-box idea about sequencing several plasma containers together to allow a cooling time for containment vessel while maintaining the fusion reaction endlessly and indefinitely. She was thinking about trying the idea, if the rest of the team didn't shoot it down.

Jim's mind was still on the spying attempt back at the lab. He was beginning to realize just how vulnerable they were. He felt that more security should be in place. He had ideas for General Wilson. He would call in the morning and see what Frank was thinking. At the same time he thought it would be good to plead leniency for the two men who had been hauled to the Stockade.

As they sat relaxing and sipping their beer, Denise noticed two men in the parking lot. They were looking over Jim's Mustang a little too closely. Jim and Denise grew silent and watched for a few seconds. It appeared as if the men were looking for a way to break into the vehicle. Jim instinctively yelled from the balcony before he thought. It might have been better to catch them red-handed in the act. They might also be spies from Houston. The two men looked up at the balcony, then ran. They were long gone in a matter of seconds.

After that incident, Denise and Jim were both more upset than ever. So, they decided to have a couple more beers before calling it a night. They wanted to be up early to get started on the tests. It was still fairly early. Jim checked his Rolex. It was only 10:30 PM. Just the same, it would be more difficult than ever to get to sleep now.

There wasn't anything in the car that spies might want but he didn't relish the idea of anyone messing with the new Mustang. It was equipped with a sensitive alarm system. It would be very difficult for anyone to do much damage without attracting a lot of attention to themselves.

Just the same, he called the office of the Hotel and told them what they had just witnessed. Within minutes a security guard on an electric golf cart appeared on the scene. Jim quickly ran down the stairs, introduced himself and pointed to the Mustang. The guard told him he would be on duty until 9:00 AM, and promised to make frequent checks the remainder of the night. It made Jim and Denise both feel a lot more secure. After their fourth beer they went on to bed. Everything appeared to be quiet the rest of the night. They both slept well and were up at 5:00 AM, ready to begin the day.

The entire team met in the lobby and agreed to stop at a nearby Denny's for breakfast. Their intention was to be at the lab by 6:30. AM. It worked out perfectly. Of course Jim picked up the tab and tipped the two waitresses $20 each.

At 6:30 AM they walked through the door of the lab. Everyone there was casting dirty looks at them. The story had obviously gone out about two of their own, spending the night in the Stockade. The reception was cold to say the least. Before they had gone ten steps, the director of the lab met them in the hallway and asked them all to join him in his office.

He introduced himself cordially and shook hands with everyone in the group personally. Once they were all crowded into his small office he offered everyone coffee and tried his best to maintain a smile. He wasn't happy, but he wasn't angry either. He demeanor might more aptly be described as terrified. He knew who the General in charge of the DOD was. He wanted no part of the DODs wrath. He made that his major point as he began.

Immediately, the three day restriction for their testing was lifted. He told the entire group they could use the facilities as long as they wished. He was at their service. If there were any issues what-so-ever to please notify him first. He would rectify any and all problems. At the end of the conversation he asked Jim personally to please see what he could do to get his two best men out of the pokey.   As the situation currently stood, they were not to be released for any reason until General Wilson gave the word, personally.

"Please call off your bulldogs!" he pleaded. "You will have the run of these facilities. We'll go out of our way to keep you people happy. You have some mighty powerful friends in Washington. We seem to be off on the wrong foot. Let's start over as friends. We're actually all on the same team, here. It isn't a competition. Our men were just curious and we're totally sorry. We apologize."

Jim placed a call to General Wilson as they sat there in the office with the director pleading his case . . .

"Hi Frank. It seems we've turned the Sandia labs upside down. I want to ask a favor. I'm certain those two guys didn't get any sensitive info on our work. They did take pictures of our setup, but we haven't brought in the entire EG vortex unit yet and we have their cellphones with all the pictures. I've deleted them.

They haven't even seen any of the really sensitive stuff. I'm wondering if you can get them out of the pokey. The director here at the lab is only concerned for his men. He's looking after them as any good supervisor should do. Can we get those guys out of the Stockade? I'd consider it a personal favor, Frank."

Frank spoke rather sternly even to Jim, his buddy, "Listen Jim, we do need to be careful here. There are factions every where that want to see you fail. They'll do whatever it takes to bring you down. It means big bucks to them. If they can steal your ideas and your work, they will. If they can sabotage you, they will. I'll tell you what. I'll get the two guys out of the Stockade but I want them both on vacation as long as you guys are there doing your tests. I don't want them anywhere near you guys or the base, and especially the lab. Pass your phone over to the director. I'll have a few words with him, then I'll call and get the guys out of the Stockade. They are officially on paid sick-leave until you people are finished with your tests. "

Jim handed the phone over to the director and within a few minutes he handed it back. His face was flushed but he was smiling big.

"General Wilson is calling the Stockade setting them free right now. Thanks for that. You shouldn't see them again as long as you are here. They will be getting a few days of R&R. They'll be happy about that and they'll probably be in Vegas before sun up tomorrow. Thanks again. I know you people are anxious to get to work, so I'll get out of your hair now. You have total run of the place. If there's anything you need, anything at all, please call on me."

Everyone left the office, nursing coffee and smiling. It was time to get to work.

As they entered the lab they noticed that two AP guards were stationed at the door of the lab. The guards even asked to see everyone's ID and security classifications. The entire team breathed a sigh of relief as they entered the lab and went to work. The additional security was a blessing.

Chapter 13: Vortex Test #5 - Tragedy

They began by examining the testing equipment. By 10:00 AM the were ready to begin with test #1 of the 5 tests they needed to do. The first test went splendidly. They repeated the same test three different times and three different ways. They were making certain that the electromagnetic pulses were calibrated to within the very thin spec-margins that were necessary. The preset calibrations were exactly correct in relation to the equations and design diagrams. Test #1 was a complete success. They moved on to test #2.

The setup for test #2 took several hours. It was 6:00 PM before the test setup was ready. The test itself would only take a few minutes. They took a vote and decided to run it, even though they had already been working for nearly 12 hours straight. Test #2 was a success.

Test #3 followed the next day and was successful by 3:00 PM. They had test #4 setup by the end of the second day. On arrival at the lab the next morning, test #4 went well but there were some worries. They worked with the equations and found a tiny error that would cause a necessary alteration to the electrogravitc vortex assembly sensors.

Gus and David went into MacGyver mode and jerry-rigged a work-around solution that allowed the team to continue. The ingenious repair job allowed test #4 to pass.

Dr. Enderman looked over their work and checked it thoroughly several times. He was amazed at their resourcefulness. He called the Gus and David's fix, a stroke of pure genius.

The #4 testing procedures alone had made the trip to Albuquerque worthwhile. It had proved a basic design flaw that might possibly cause trouble in the working model of the finished reactor. It essentially meant that they would need to overhaul the vortex assembly to different specs on their return to Los Alamos but now they could see the error and they knew how to remedy it.

Test #5 would be a heat test. It would need to be done in a different lab facility about two blocks away. It meant moving to the TA-III heat-testing facility. This was to be a test of the housing assembly of the vortex containment housing. There were debates as to whether or not this test should be conducted.

There was a vague fear that the jerry-rigging that had been done the day before might not hold. It might blow under the stress of too much heat. After about two hours of debating and arguing, it was decided that a simple, easy test should be done, first. The objective was to hopefully discover the varied tolerances of the materials they were using. It was a calculated risk and they wrote it up that way. They knew it might be dangerous, but they were in hopes it would save time. They might discover the degree of flaws and it would help them qualify their corrections as they rebuilt the new version of the housing assembly. It might save them from making several more trips back to Albuquerque for more tests, later. They decided to go ahead with test #5. It took three hours to set it up.

Dr. Enderman took charge and signed-off on the decision to do the test. He took full responsibility. It meant that he would be in a precarious position. He would need to be manning the controls as the test was run. All the others in the group would move away to safer confines. At 4:00 PM they were ready to run the test . . .

Dr. Enderman took his position at the control panel. The entire team moved to a safe position almost 50 yards away. They all grabbed cokes and snacks at a vending area that was a considerable distance from the actual lab and control panel.

They heard the equipment start up. There was a whirring sound, then a muffled bump sound. Smoke poured from the door of the control room and Dr. Enderman stumbled out the door. Smoke was coming off of his smock as he fell to the ground. Everyone but Jim rushed to his side.

Jim pulled his cellphone and called the Kirtland AFB emergency number explaining where they were and what had just happened. Then, he too went to Dr. Enderman's side. It didn't look good. Doc Enderman wasn't burned all that bad, but the concussion of the blast was apparently much worse than it had sounded to the team. All they had heard was a bump or plump sound, but the lab was well insulated. It had been designed to contain a horrific explosion.

Doc Enderman's eyes were open but glazed over. He appeared to be dead. A panic fell over the team. Both girls were losing it, crying almost hysterically. Dave and Gus weren't doing much better. They both stood there wringing their hands and shaking their heads.

Jim stooped to one knee at Dr. Endermans side and checked his breathing. "Nothing!" he said. "He isn't breathing and there's no pulse!" They could hear sirens in the distance but the EMC was still a minute or two away.

"We need to do CPR and mouth to mouth right away. Raise his feet to keep him from going into shock." He immediately began doing CPR and Denise gained enough composure to begin mouth to mouth. They were not getting a response of any kind.

Seconds later Audrey had directed the EMS wagon over to where they were. Other people came running from the various labs and offices. Three men went into the control room to secure the heat lab.

Dr. Enderman was loaded onto a gurney as the EMS team tried desperately to revive him. One of the EMS team looked at Jim and shook his head negatively saying, "it doesn't look good. How long has he been unconscious?"

By this time, it had been 8 or 10 minutes. "8 to 10 minutes." Jim replied.

The medic shook his head again. "If we can't bring him around in the next few minutes, I fear the worst." That's all he said as he closed the doors of the wagon. It sped away for the base hospital with sirens crying and lights flashing.

Jim spoke with the lab director and explained that they would all be going to the hospital. There would be no further testing but they would want to examine everything and take pictures for their reports. As soon as Dr. Enderman was able, they would be going back to Los Alamos. They shook hands. The director was very sympathetic.

Soon after that they all piled into the two different vehicles and headed to the hospital.

There was a very tense wait at the hospital. The doctors and nurses were all mum. That's always a bad sign.

Finally a middle-aged doctor in a smock came out with both hands clasped in front of him and a sad look on his face.

"We did all we could to to revive him. He was actually dead at the scene. He had a severe concussion but we also think it was just too much for his heart. He had a massive simultaneous heart attack too."

Gloom set in at that point. The entire team felt responsible. They weren't sure yet what had gone wrong but they feared it was the jerry-rigged vortex assemby fix they had done. Gus and David felt horrible about that.

Later they all went back to the heat lab and examined the damage. They took pictures and wrote reports. As it turned out, the jerry-rigged Vortex assembly was still in tact. That wasn't what had caused the problem. Something unknown had caused the vortex enclosure to blow out at one end. Apparently it was a defect in the materials they had used in the assembly housing. Under the pressure and the heat, it just blew. Everyone was shocked as they looked around inside the lab and control room. It had been a rather large explosion. The insulation and design of the building had done its job. It had contained the explosion very well.

Everyone felt miserable. This accident might cause big changes in the project. It would definitely slow them down for awhile.

They would probably all be going back to DC for the funeral and there would be investigations by the project brass and staff. Everyone would want to know what happened and why. The future of the entire project was suddenly in jeopardy.

Jim spoke with the hospital administrators and made arrangements to have Dr. Enderman's body flown back to Washington DC. Then he had the unenviable task of calling General Wilson to tell him that one of his best friends was dead.

General Wilson was very upset, but he was a superb leader, or he wouldn't be sitting where he was sitting. It was wonderful the way he reassured everyone on the team that the project would continue.

"We'll figure out what went wrong. We'll fix it and move on. Don't even think about giving up. Doc Enderman wouldn't have it that way and neither will we.

Jim put in a call to have the Learjet flown down. They would all fly back to Washington for the funeral. It was a matter of respect.

They loaded what was left of their reactor assembly back on the truck and the two employees from Los Alamos drove it back.

The team stayed at the hotel for the night. They would leave their cars in Albuquerque and fly out the next day on the Learjet, to be in Washington for the funeral.

"It's just a slow-down in our project." Jim reassured everyone. "We'll figure out what happened, fix it and move on. We will complete this project. Failure is not an option!"

Denise began looking at all the photos of the damage along with Gus, David and Aubrey. Until they could explain what had happened, they wouldn't know how to fix it.

Finally after a couple of hours examining and reexamining the photographs, Gus saw it. A simple screw had worked loose inside the assembly. That's all it was. A man had died because a screw had worked loose. That made it all the more depressing. Denise and Aubrey were both crying off and on.

The next morning they drove both cars to the International Airport and put them in the long term parking area, then they all boarded the Learjet and flew back to Washington. Everyone on the team dreaded the fact that they would need to go before the staff review and explain what had happened. A faulty screw!

Chapter 14: Moving on . . .

The funeral was totally depressing. Dr. Enderman was laid to rest amidst sobs of grieving friends and family. But finally it was behind them. Denise's brother Eddy showed up to support his sister. She introduced him to everyone. Jim now knew who her inside source was, at the DOD. He considered it a good thing and vowed to help Eddy keep it secret.

Then there was the staff review. Jim and the entire team was called into a high-level Pentagon meeting. Denise took the reigns and showed the Pentagon brass all the pictures, explaining how a simple screw had done all the damage. The photos proved it when they were viewed in the proper sequence. She was adamant that it was an easy fix. She put everyone at ease that the project was not in peril. "We can and we will accomplish this." She said.

Dr. Enderman's autopsy report made everyone feel a lot better. He had been living with a seriously bad heart and had kept it secret from everyone. The autopsy showed that it wasn't the small explosion that had killed him. It was entirely his heart. He was facing a very limited time left on earth, and he knew it. The sudden shock and trauma of the explosion had set off the heart attack that killed him. He had died doing what he loved.

Dr.Enderman himself had signed-off on the test. It was documented. He was well respected. If he thought the test was safe and if he signed his initials to that effect, it meant there was nothing more to discuss. An accident had happened. That was it.

To everyone's surprise, General Frank Wilson made a motion that Dr. Enderman's position needed to be filled, but he had all the confidence in the World that Denise was more than capable of becoming the lead scientists on the project.

The General took a vote among the DOD staff members and made it official. No one objected. Denise would take Dr. Enderman's job and there wouldn't need to be an additional replacement.

When the meeting was over, there were a lot of handshakes, kudos and encouragement. Everyone wished great success for the team. They were free to return to Los Alamos and resume their work.

A considerable amount of celebrating was occurring as the Learjet 45XR lifted off from Reagan National Airport, bound for Albuquerque.

Denise was still in shock that they had made her the head of the team. It also meant an extra $50,000 per year would be added to her salary. Everyone on the team was happy for her. They all agreed that she had been the team's primary leader since the beginning back at MIT. She had hand-picked each individual from a choice of hundreds of MIT grads and she alone had scored the Starlight Project positions for all of them. There was no dispute as to who should lead.

By the next day they were back at Los Alamos rebuilding the reactor assembly to all the new specs.

It might take approximately another six weeks to rebuild the new electrogravitic housing assembly. They all fell into an eager routine of daily work. Though they had vowed to not work on Saturdays and Sundays it often happened that one of the team members would wonder down to the lab on a day off, only to find that two or three of the others were there as well. That's the way it goes when you love your work.

Denise was notorious for pulling up a stool beside the new assembly unit on a Saturday or Sunday and she would sit there for hours sometimes, just looking at it, while rearranging everything mentally, as she sipped a coke. To her it was like a game of chess. She needed to mentally visualize every possible thing that might happen; every potential hazard and every potential outcome. She would often take the Mustang if Jim just wanted a little 'me' time. She claimed she could think clearer when there were no people around to break her concentration.

Within four weeks the new assembly was almost ready for testing again. The second rebuild went a lot faster because they didn't have to spend as much time thinking and they had the diagrams on the computer to lead the way. The fact that Denise had spent hours on the weekends just looking at the thing and thinking, made the work accelerate much faster when everyone was there.

By then she knew exactly how to overcome every problem before it arose. She had already thought it through. If the group challenged or disputed her ideas, she would always have ready answers that they couldn't dismiss. In other words she stayed about a half mile mentally-ahead of the others and she was almost always correct in her thinking. By Wednesday of the fourth week, the new assembly was ready once again to go to Albuquerque for the same five difficult tests that couldn't be done at Los Alamos. They would need to do those same five tests again. This time, they were a little more confident the unit would pass all five. Everything had been checked and rechecked a hundred times.

They called the director at Sandia and were given carte blanche access, this time. The tests were set to begin on the following Monday.

On Sunday night, Eddy called Denise out of the blue, "I see you are headed back to Sandia tomorrow. I saw something on the DOD computers today that means that you need to be especially careful. **If your unit passes all the tests, watch out!** I think there may be a sinister plan underway. I'm not sure who they are or what they are planning. It was only hinted, using coded language, but be on-guard and expect big trouble."

Denise found that piece of news unsettling but she thanked him and asked him to keep her informed. It gave her great mind good cause to stir with a very special idea. It would and could turn out to be a very inexpensive form of insurance.

When Monday arrived, the Los Alamos convoy set out for Albuquerque again. This time Denise would be manning the control panel on the worrisome heat test #5. They were all a little anxious about it, but they also knew they were ready.

It turned out great. The new assembly passed all five tests without any problems. They were finished with all the tests by 2:00 PM Wednesday. Everyone was thrilled beyond words but they remained secretive about the test results. They all felt it would be wiser if they didn't celebrate in front of the regular lab staff at the Sandia laboratories. The agreed as a group that if word went out that they had been successful, it might lead to unnecessary troubles. Mum was the word and they all acted as if they weren't happy about the tests when prying ears were listening. They all knew however that a major milestone had just been reached and they were silently exuberant.

Jim did fire off a special e-mail to General Frank Wilson:

"The vortex assembly is finished. It passed all five tests with no flaws what-so-ever. Next step, controlled and sustained fusion! ~ Jim

Now, they could begin putting everything together back in Los Alamos. The next phase would be to put the entire reactor together and actually begin creating fusion reaction experiments. Los Alamos had a facility where they could do those tests. This might be their final trip to Albuquerque for awhile.

On their final day in Albuquerque the entire group including the two who drove the military transport truck, celebrated by going out on the town. Jim treated everyone to dinner on him and a tour of Albuquerque's night life. They celebrated until around midnight at the Sandia Resort and Casino.

Gus and David were now (or soon to be) young and wealthy playboy types. They would occasionally score a cute babe for the week-end and it would be over on Monday morning. Neither was interested in finding a long term relationship. They both enjoyed the freedoms of bachelorhood.

Everyone had always hoped that Audrey Reynolds would connect with either Gus or David, but the chemistry just wasn't there for any of them. They were all great friends but the love magic just wasn't there. They had tried and they were all too close, as friends. They had partied together and had become too drunk, then ended up sleeping together, only to regret it the next day. She wasn't a love-match for either of them and she wasn't attracted to either of them that way. Though she was devastatingly beautiful, she was dreadfully lonely. Her love-life had been in the pits. It often showed in her personality that she needed someone on a permanent basis.

As the group partied and celebrated at the Sandia Resort and Casino, Audrey was approached by an amiable and dashing guy who immediately held her spellbound. They connected, drank, laughed, dance and enjoyed each other's company. His name was Enrique T. Smith.

The name was so phony sounding it was almost comical, but apparently it was his very real name. He was an all American type, of mixed Hispanic ancestry. The hot Latino blood was there. He was tall, dark and handsome, with coal black irises and a million dollar smile. His English was impeccable with only a sexy hint of an accent, even though he looked more Hispanic than Anglo. He was the type of guy that any girl would quickly fall for. First of all, he was a nice guy. He was thoughtful and very polite. Enrique had a permanent 5 o'clock shadow that gave him a slightly rebellious look; and those eyes were mesmerizing, mischievous and devilish. Audrey was immediately smitten. She brought him over to the table and introduced him to everyone there. Before the night was over, he had been accepted by everyone. He was an easy guy to like. Everyone liked him. Denise liked him and Jim liked him too.

Enrique was the great, great grandson of an old Mexican hero Emeliano Zapata who had been a major revolutionary figure in the Mexican Revolution. He was very proud of his Mexican heritage. The Mexican side of his family had been very wealthy and he was an heir to a lot of it.

Audrey left the Casino with him at the end of the night. Everyone hoped it would only be a one-night fling. He brought her back to the Hotel on Friday morning and kissed her goodbye. Her eyes were still spinning like hubcaps after a collision, and her heart was doing back-flips. It was fairly obvious that apparently, Enrique had rung the bells at St. Pedro. For her, it was a little more than a one night fling. Enrique seemed to be floating on a cloud as well. They made a lovely couple and the chemistry was definitely there. Enrique wanted to drive her back to Los Alamos, but he had obligations with the family business. He couldn't leave. He lived in Albuquerque and Audrey was on her way back to Los Alamos.

She had not told him anything about the project. As far as he knew, she was simply living and working in Los Alamos in a Department store. They kissed and said goodbye as the team readied to return to the labs of Los Alamos. He promised that he would visit within a few weeks. She gave him her cell number and the address on Arizona avenue.

The team packed and prepared to return home. They knew they had been closely watched nearly every second of every day they were there, but nothing out of the ordinary had happened. There were AP guards posted with their equipment 24 hours a day at the lab. There hadn't been any threats at all. They simply did not think of themselves as important enough for anyone to worry about, yet. Though they were working on a potentially world-changing project; up to this point they hadn't done anything spectacular that should draw any special attention.

Just the same, it was obvious now that a lot of big money people would like to see them fail. The closer they came to success, the more danger they would be in. Project Starshine wasn't as top secret as they thought and assumed it was. The project was quickly becoming a threat to some very big and powerful corporations . . .

Chapter 15: Ambush!

On Friday morning by 11:30 AM, the truck was loaded with all their equipment and their baby, the perfectly working new electromagnetic reactor assembly, including all the equipment they had brought with them to Albuquerque.

Denise worried as she remembered the warning call her brother had made. He had said:

"*If your unit passes all the tests, watch out.*"

That warning phrase was playing over and over in her head. She knew the reactor inside out. Within the vortex assembly, there was a very small sensor that was crucial to the calibration of the electromagnetic repulsive forces. It reversed the polarities and focused the vortex inward or outward. The assembly would fail to work without it. She had designed it and she was the only person in the World who completely understood how necessary it was and how it worked. It was about the size of a quarter. She removed it and shoved it into the pocket of her jeans, smiling inwardly.

The convoy set out for Los Alamos, just 90 miles away. Audrey, Gus and David were in the Subaru, in front of the truck. Jim and Denise traveled behind the truck. The girls talked on the cellphones, but they were both being much more careful than before.

For the most part, Audrey talked about Enrique as Denise listened with her eyes rolling back into her head. She would grin at Jim and hold the phone away from her ear as she pretended to listen, while smiling and shrugging her shoulders. She listened to Audrey's banter about Enrique, because that's what friends do. It was mostly boring mush.

They all realized that they were a lot more vulnerable now that the reactor unit was essentially finished. They didn't talk about it at all. Their working model had passed all the tests. It was suddenly a potentially valuable invention.

As they traveled in the convoy they went a little slower than the other traffic. Denise noticed a black car about a half mile behind them that slowed when they slowed and picked-up speed as they accelerated. She mentioned it to Jim and he began watching it as well. A suspicious chopper flew in wide circles above. Were they being paranoid or was something about to happen?

Shortly after they had turned onto hi-way 502 just a few miles from Los Alamos, what appeared to be a hi-way patrol car pulled up in front of the truck and beside David's car with a siren chirping and lights flashing. He motioned for all three vehicles to pull over to the side of the road, which they did. Instantly they were all surrounded by black SUVs. The hi-way patrol vehicle disappeared quickly over the horizon. Apparently, it had not been an authentic patrol car, but It had performed its part of the operation and was gone. There was no other traffic on the lonely road.

The men inside the SUVs wore hoods and they were brandishing rapid fire weapons. As both cars and the truck pulled to a stop at the side of the road a black SUV pulled up behind Jim and Denise. A man with a black hood over his head, motioned for them to get out of the car as he pointed an AK47 at Jim's head. At the same time, the driver of the truck was also escorted to the back of the truck and ordered to open the tailgate. As the tailgate swung open, another very large truck seemed to come from nowhere exactly on cue and pulled along side. A total of five men wearing hoods unloaded all the equipment from one truck and onto the other. They were quickly gone. It happened very fast.

It was over and done in under four minutes. They were gone with the new electrogravitic assembly, the reactor and all of the other equipment. It was a very well orchestrated robbery, executed perfectly by experienced professionals. The entire caper was well planned and carried out with lightening speed.

The Starshine team all stood around the now empty Los Alamos transport truck. They were in a total state of confusion and shock. All their work had just disappeared in front of their eyes and in broad daylight!

Denise gleamed a little more than seemed appropriate. She reached into the pocket of her jeans with a sly smile. "Who can tell me what this is?" She was asking the whole group as they fretted over the loss.

Everyone on the team looked at her curiously. Gus looked at it and laughed out loud! Then he hugged her neck! "That's why you are the team leader!" He said it with such a smile and with such joy that everyone immediately knew she had saved the day in some mysterious way.

Gus took over for her. "It's the sensor! The vortex assembly will probably blow up without it. Whoever stole the vortex assembly had better know what they are doing, even if they are just trying to test it. It'll blow them sky high without that sensor. The odds are, they won't know. It's the key ingredient to the success or failure of the project. You're a genius, Denise! Pure and absolute genius!"

Jim, always the cool head was also smiling at this bit of good news, but the situation was still dire. It was just not as dire as it might have been without Denise's little insurance policy. He immediately punched the speed dial on his cell phone to General Wilson at the DOD. It rang twice before Frank came on the line, "Hey Jimmy my man, what's up?"

Jim spoke softly but with a deeper seriousness than usual, "Frank, we have another serious problem out here in Los Alamos. We've just been hijacked about 8 miles from the Los Alamos base. We were pulled over by a fake state trooper, then ambushed by a crack squad of goons with hoods and machine guns. The short version of the story is that they have stolen everything and disappeared.

The vortex assembly passed all the tests with flying colors at Sandia. We were on our way back to Los Alamos with it. It was ready to go. We might have had a positive fusion reaction within another month! Now we are at square-one and someone very big and very professional has disappeared with the entire vortex assembly. We are currently sitting on the side of hi-way 502 about 8 miles from Los Alamos. It happened just minutes ago. I guess we'll be heading on back to Los Alamos. There's nothing much we can do on this end. That's the bad news.

Now for the good news! Denise; genius that she is, had removed a crucial sensor as an insurance policy. They won't know it belongs on the unit. If they try to do anything with the reactor, even test it . . . it will blow sky high!

Do you have any ideas, Frank? What can be done? Is there anything you want us to do to help from this end?"

Lieutenant General Frank Wilson was immediately incensed. "The Hell you say! They got everything? Are you folks okay? Did anyone get hurt?"

"No physical damage, Frank; but I'm mad enough to have a coronary. Who would do this?"

"Tell Denise she has just made me proud. I knew she was the right one to head up the team. I doubt if Doc Enderman would have thought to sabotage his own work that way. Brilliant!

I have suspicions about who the culprits might be, but it could be several different outfits, Jim. There are a lot of people who want to see you people fail. I wonder if they know who they're messing with. I'm going to have some heads on platter over this. Jim, don't worry, we'll put the CIA, the FBI and the Marines on it if necessary. We will get that reactor back and there will be prosecutions. We'll nail 'em Jim; don't you worry about that.

You guys go on back to Los Alamos and stay close by the base. I don't want anything to happen to any of you. Whoever has done this isn't afraid to play rough. Keep yourselves safe. We need each one of you on this project.

I'll go to work on it from this end. I'll get back to you as soon as I know something. Keep your chin up buddy. I'm blaming myself here. We should have provided better security for you on the road. I guess they caught me sleeping at the wheel. I'm awake now! You can bet on it. Heads will roll! We'll get 'em Jim. Take a break and be patient. I'll be in touch as soon as I have news. So long, buddy."

Jim X bid a farewell to his good friend. "Bye Frank. Thanks. " They both hit the disconnect.

Everyone was crowded around Jim, trying to decide what to do next.

"Well guys, there's nothing much we can do. Frank will have the CIA and half the alphabet looking for those guys within the next 10 minutes. We may as well go back to the base and sit tight." Jim was rubbing his hands together and shaking his head. Then he hugged Denise and held her close. "Thank God for you and your marvelous mind! You may have saved the day."

The convoy went on to Los Alamos. It was depressing for all of them but it would have been much worse without Denise's wise thinking.

Chapter 16: Retreat, Regroup & Rebound.

The team arrived back at the house on Arizona avenue around 4:00 PM. Jim supplied the answer to all the gloom and doom. He pulled five ribeye steaks from the freezer and dug around in the bar cabinets until he found a bottle of Jose Cuevo Gold. Along with that, he produced all the ingredients for Margaritas. They would grill steaks and sip Margaritas by the pool until the pain of it all had gone away. They had everything they needed. The party was on!

They hadn't actually relaxed at all since they had arrived in Los Alamos almost three months earlier. It was a great idea. They all sat around the pool in bathing suits, drank Margaritas, swam and tossed ideas back and forth about what to do next.

By 10:00 PM it was decided. They would return to the lab on Monday and begin building reactor assembly #3. No one on the team wanted to sit on their hands waiting and hoping for a return of the stolen reactor. The odds were especially good that whoever had it, was about to blow it up, along with anyone who might be within 100 feet of it. They worried about that. Innocent people could get hurt. But, they hoped it would be recovered before that could happen.

The party ended by 11:00 PM and they all went to bed. Sunday was spent about the same way. They grilled burgers and sipped a few beers. They were all alike. Each was anxious to begin working on the next unit. No sense crying over spilled milk. The only thing to do was keep moving forward.

So, on Monday they were all back to work at the lab, building a third version of the project. They had done it so many times now it was easy. They were 40% finished by the end of the week.

On Friday after work, Denise got an encrypted e-mail from Eddy. She was anxious to decipher it. She was almost certain that Eddy might have a lead on the stolen equipment.

Sure enough, he did. It took her 30 minutes to decipher the whole message on her laptop. She became shocked as she read it in private . . .

"Sis: *There is a mole inside the DOD. I have determined who it is but I will get into serious trouble if I tell anyone. I'm breaking rules to gain the knowledge I have. As you know, my girlfriend Rachel does computer security for the NSA. She has been breaking a few of the same rules over there. There is a mole inside the NSA too. Her mole and my mole think they are exchanging encrypted messages but we have the encryption key. We're breaking all the rules and reading all their messages because we are more concerned for the safety of you and your people, than we are for ourselves. If any of the higher-ups find out, we could get into serious trouble for doing what we are doing.*

Anyway, we've discovered that there is an outfit in Houston, a criminal element within the big oil companies. They are wanting to shut you people down completely. They're talking about eliminating you! They're discussing different ways to discourage you from pursuing the fusion experiments any further. They really don't seem to want to hurt anyone physically, unless it is the only way to stop you. They're looking for ways to discredit you so that the DOD will drop you like a hot potato.

THE BOTTOM LINE IS THIS:

Mr.X (Jim) needs to come up with a way to get the brass at the DOD looking at their own people. They need to find the mole, then squeeze the truth out of him. He holds the secret to where the stolen reactor is.

What I'm about to tell you could get me 20 years in prison. The mole inside the DOD is an Army Captain named Dugger! I hope this will help you. Please, please, please protect me as your source. If they find out how I've learned all this, Rachel and I both will go to jail. We've committed some very serious security breaches ourselves. The mole's contact inside the NSA is Thomas Williams. He is a civilian.

Good luck and stay safe! Love you sis. ~ Eddy

Denise was reluctant to show the message to Jim. She thought about it for a long time. Her brother was sticking his neck in a noose for her. She had to protect him. She decided to play it cool. She grabbed two beers, hugged Jim around the neck, planted a wet kiss on his lips and whispered that they needed to talk outside. Then she led him to the solitude of the pool and they took a seat at a wrought iron table under an umbrella.

She rubbed her forehead nervously before she began. "Jim, do you know a Captain Dugger at the DOD?" She asked.

"Yes! He was assigned to help me pick the project team and he set us up, here at Los Alamos. Why?"

"Do you trust him?" She asked with a hint of suspicion in her eye. In a way she was letting him know the man wasn't to be trusted, but she was trying not to say it directly.

Jim grinned, then laughed. "I know you pretty well by now Denise! I can tell from that look you just gave me that you know something I don't know. Spit it out. What's up?"

"I'm being a little coy for a good reason, Jim. How well do you know Dugger? Are you sure he can be trusted?" She was seriously worried about implicating her brother.

Jim had already figured everything out. "Your brother knows something that will get him thrown in prison. Am I right?"

She had to laugh and look away. "Boy, I'd make a lousy spy wouldn't I. You read me like a book. Jim, there's something big. It's really big, but Eddy has his neck in a noose. No, not just a noose; it's a Guillotine! We can't allow him to be implicated as our source, okay?

He's doing it because he loves his sister and he wants this team to succeed. We have to find a way to keep him out of it completely, but Dugger is definitely a mole. He knows everything we do and there's a mole at the NSA too. Eddy's girlfriend works over there. Together, Eddy and Rachel have broken every rule there is, to get at the story. The two moles are involved with some outfit in Houston, a criminal wing of the big oil companies. They have the stolen reactor and all of our equipment.

They're connected in some way to the big oil companies who want us . . . in Eddy's exact words: . . . 'eliminated'. I'm as afraid for my brother as I am for us. Now look me straight in the eye, Jim. This is important!

What I have to say next, is important to you and me. If my brother gets nabbed for helping me . . .or us . . . on this, I'll walk away from this project and I'll walk away from you. I think you know how much I've grown to love you over these past few months. We simply must find a way to trap the mole without implicating Eddy. Do you understand my point and my worry here?"

"Yes, yes, yes." Jim was on the same page with Denise immediately. "I completely understand and I will never betray Eddy. That's a promise. But, how do we do it? We've simply got to alert Frank about the mole and it's imperative that we catch these people. Do you have any ideas? I mean, I know Frank Wilson like a brother. I don't think he would ever allow anyone to prosecute Eddy or Eddy's girlfriend for being our heroes, but I agree; we can't and shouldn't take that chance. We need to find a better way to either trap the mole ourselves, or find a better way to alert Frank that there's a skunk in his office." Jim rested both elbows on the table with his chin in his hands as his own great mind turned it over and over.

Denise was perplexed as well. "How often do you talk to this Dugger fellow?"

"Not very often." Jim responded. "He helped at the very beginning but once we had setup the team and moved out here, I haven't had any communications with him. He works in the DOD office with General Wilson. Why? What are you thinking?"

She tossed Jim her devil smile. "I'm thinking of a way to trick him into revealing himself to General Wilson. We might pull it off without saying much at all. What if we feed Dugger some false info that would cause him to trap himself, in front of General Wilson? Can you think of any way we can make that happen?"

Jim tossed that question around in his mind a few seconds. "The secret to that, might be in the messages he's sending or why not just tell Frank that I'm uneasy about something Dugger has said to me and have Wilson check him out? Wilson is sharp as a tack. If we put him on the trail, he'll smell the skunk. He doesn't sit where he sits, by being stupid."

"Here it is! I've got it!" Jim was grinning big, now.

"The Ambush! They knew the vortex assembly had passed all the tests. They knew it was ready! I sent an e-mail to Frank Wilson as the tests were completed. Dugger had to have intercepted it.   Dugger is the only one who would have told them the reactor was complete and finished. He told them when to setup the ambush, since we know Frank wouldn't do that.

Dugger is the only other person besides Frank who could have seen the e-mail. We didn't tell anyone at Sandia how our tests went. We were top secret about everything there. They didn't know if we had succeeded or not. It proves that Dugger is the skunk without implicating Eddy. How does that sound? All I need to do is imply that I'm suspicious of Dugger and Frank will put him over the coals."

Denise thought long and hard before she spoke. "It does sound good. I think it's the best plan we've got. Let's do it!"

Within seconds Jim had speed dialed Frank Wilson and waited for him to answer. "Jim! How's it going out there?"

"We're recovering. The team started building the third reactor and they think it'll be ready for testing in another week or two. Are you having any luck at tracking down the stolen prototype?"

Frank responded. "No! Nothing has jumped out at us. We suspect some people in Houston, but we can't move in on them without evidence and we don't really have anything but suspicion."

Jim was cautious, "Frank, we've done a lot of thinking. The people who ambushed us, knew too much. They knew that the equipment had passed all the tests or they would have waited to steal it. They wanted a working version. That's why they were quiet until then.

I'm about to suggest a delicate and suspicious situation with this, Frank. Think carefully about what you hear me say next.

Only six people were supposed to know the assembly had passed the tests; the five of us . . . and you. I sent an e-mail to your computer within an hour after the tests were complete. Did you see that e-mail?"

"Yes, I saw it; but I didn't check my email until around 5:00 PM. I always check the e-mail as I head home for the day. What time did you send the e-mail, Jim?" The General's mind was already beginning to suspect Dugger. It was obvious.

"I sent it at 2:00 PM, your time, Frank. Is there anyone there in the office who could have intercepted that e-mail? If there is, I think we can assume that you have a mole, there in your office. Are you with me? We know you wouldn't setup an ambush and there is no one out here who knew the reactor was two-thumbs-up as an essentially completed project. I think you need to do some detective work there in the office. Something smells pretty bad." Jim looked at Denise with oval eyes and a lump in his throat.

"Dugger! Dugger was here all day. It has to be him. It has to be! That's some pretty good detective work from your end, Jim. I'll smoke the SOB out. He is on the carpet as of now. I'll call you back when I know more. Thanks for the tip."

"Frank, keep in mind that if Dugger passed the news, he will also know who has the reactor and where they're keeping it. Be smart. Nail him on the e-mail first. Get him to admit to something small. Then hold his hands over the fire until he tells you who and where. I'm wishing you luck old buddy. We want that reactor assembly back and we want the people who have it, in jail . . .

forever! Thanks Frank. Good luck!" They both hung up at the same time.

After he hung up he looked at Denise like a puppy begging for a bone. "How did I do?"

"It sounded great to me. Except there's still a small nagging problem. There is a mole inside the NSA too. His name is Thomas Williams. He is also a big part of this nightmare. We can only hope that Dugger will give him up. If he doesn't, we'll need to find a way to put Frank on his trail too. I can't wait to see how all this turns out!"

Chapter 17: Hunting Skunks - 101!

Most of the team enjoyed a relaxed week-end sitting around the pool at the house on Arizona avenue. Enrique had called. He was missing and longing for Audrey. She invited him to come to Los Alamos for the week end. They partied and played in the pool along with most of the team. Enrique's unique personality had won everyone over. They were all beginning to see him as a friend. Audrey was totally smitten with him. The more time they spent together, the more the two appeared to be falling in love. Everyone was happy for Audrey. Her personality seemed to change when Enrique was around. She became a bubbly bundle of laughter and joy.

Occasionally, someone in the group might slip and say something sensitive about the project, then the others would cover the comment, or change the subject if Enrique was within earshot. He didn't seem interested in any of that. His focus was entirely on Audrey.

Denise, as she so often liked to do, took the Mustang over to the lab and sat, turning every part of the reactor operation around in her mind. She felt that an epiphany was trying to emerge.

There was something interesting about the sequenced fusion plasma containment idea she had come-up with earlier in the project. At that point in time it had been like putting the cart before the horse. She needed to visualize the completed vortex assembly before she could follow-up on the more advanced sequencing idea, possibly using several plasma containment vessels. The fact that they had managed to create a working model of the electrogravitic assembly meant that she could now resume tossing the sequencing problems and ideas around. The sequencing idea seemed impractical at first glance, but as she continued thinking, it began to make sense in little short flashes.

The objective would be to sustain the delicate balance of the plasma reaction while switching. This might allow a cool down time for the various parts of the reactor that would tend to overheat. The fusion reaction would and could be self sustaining as long as it was supplied the materials it needed to fuse.

180,000,032 F. degrees is an almost incomprehensible degree of heating. It was difficult to visualize exactly how hot that could be. There was no metal on the planet that wouldn't melt and completely vaporize at such temperatures. She still wasn't certain that a sequenced vortex assembly would work well enough to protect the housing and sustain the reaction too. It was only an experiment and a hope, at best. She needed to get her IQ and imagination outside-the-box a little more.

An epiphany was trying to emerge . . . but it wasn't completely clear in her mind yet. She would need to keep thinking about it. Quite a lot was riding on the electrogravitic housing. If it worked then maybe nothing else would need to be done. If it only partially worked they might try some different idea.

On Monday afternoon, they were all at the lab working on the new reactor when the phone rang. Jim went outside with the phone so he could hear a little better. He knew it was General Frank Wilson.

General Wilson was laughing as Jim picked up and connected, "You were right big Jim! Dugger was definitely a mole. We caught him red-handed. He admitted to reading the e-mail and that led us to pressure him for the truth. It was like pulling teeth at first, but as he began to realize we had him dead to rights, and he was headed for Leavenworth, the guy started giving up names and places, bargaining for an easier sentence.

It turns out that Dugger was only a small cog in a much bigger operation. We found encrypted e-mails between him and another mole over at the NSA; a guy by the name of Thomas Williams. They'll both be bunking together at Leavenworth. They were both feeding all the info they could gather, to an outfit in Houston that is obviously on the payroll of some of the big oil people there. The long and short of it is this; we have found the reactor! It is now on its way back to Los Alamos and should arrive by tomorrow.

It turns out that they didn't want to use it, they just didn't want the World to have it. The oil companies are running scared. They don't want to see any big energy breakthroughs like that, occur. It would be a death knell to their multi-billion dollar cash cow.

We have arrested all the people who were involved in taking the reactor assembly from you out on hi-way 502. It turns out that they were a rather large criminal organization, for hire; like a small Mafia organization. They were paid really well by some unknown entity to take the reactor. As we (the FBI and the CIA) rounded them up, it turns out there were 24 guys in that operation alone. We have all but two of them jail. Two of their top people went on the run, but we will get them eventually. All of those people will stand trial. They will be dealt with very harshly by the federal government. What they did was a federal crime.

We do have leads as to which oil companies were paying them, but all that is still pretty sketchy. It may never be known. The little people in the organization weren't told who was paying them. It was all very secret. Only the top people knew, and they got away. It's is on-going and probably will be, for the next several years. If we can ever get the goods on which oil companies and which specific individuals were responsible within those companies; we will prosecute them too. So far, we are not getting very far with that. The real culprits are fairly well protected.

Anyway, I thought you and the team would be happy to know that we found everything they took from you. It was all intact at a large warehouse in Houston. It's already on the road, under armed guard and on its way back to you."

Jim was practically speechless he was so happy. "That's terrific news Frank! We had given up on ever seeing any of it again. The team has already begun working on version #3 and it is nearly ready for testing. We'll probably keep working on that until it's finished. Denise is playing with an out-of-the-box idea that we may be able to sustain the fusion process longer if they can find a way to sequence the plasma reactions back and forth between two (or more) different plasma containment vessels. It's just an idea she's tossing around but it sounds practical. By switching like that it might allow a little cool-down time, like tossing a hot potato back and forth between two or more hands. As it stands, we'll take it as it comes. First we'll probably try to sustain a reaction with just one containment vessel. As we see how successful that turns out, we'll try several different ideas until we succeed. We would like to see sustained fusion for long durations as opposed to the millisecond experiments that have been performed in the past."

The General responded. "Those are some great ideas Denise is coming-up with. It's sounds practical! I don't think we could have found better people for this project. It's a near impossible dream, but if you guys succeed, it can and will save the planet. You folks keep up the good work and keep me posted. We are here for you if you need anything. You have the complete U.S military behind you. I'll get off of here for now and hang up. Happy hour is calling my name. A military transport truck will arrive tomorrow with the stolen reactor. Thanks for the tip about Dugger. We're glad to be rid of him. He is in the Stockade now awaiting a Court Martial. He'll be doing time Leavenworth. See ya buddy!"

"Bye Frank!" Jim clicked the cellphone off and then let out a war-hoop that could be heard all over the lab. Everyone came running. They thought something was wrong. He was beside himself with happiness as he explained everything that Frank had just told him. The entire team was just as happy about it as he was.

It was nearly quitting time. Jim was so happy that he decided to celebrate in a first class way. He took the entire team out to the Blue Window Restaurant. It was known to the locals as the best in town. Besides the meal and other drinks, he ordered a bottle of the restaurant's best champagne and they all toasted the project with a bottle of Dom Perignon 2004.

Of course the discussion at the table was all shop talk. That's virtually all they ever talked about. To each of them, it was their passion in life. The project was all that interested any of them. Denise seized the moment to plan and plot their immediate strategies.

They voted and everyone decided to continue building reactor #3 to be used as a backup spare. But they would also begin to prepare a complete compact reactor simultaneously. They would use the newly tested electrogravitic assembly #2 in the soon to be returned reactor. If everything was to go as expected, they hoped to have an completed reactor ready for testing in 8 weeks or less. They were now far ahead of schedule.

At the onset, as the project began, the hope was to produce a working sustained-fusion reactor within 4 years. If they were lucky and if everything continued at the current pace, without problems, they were thinking they might have a working accelerator in under a year. That alone was worth celebrating, but there was still a lot of work in front of them. To date, they had only spent 1.9 million of the $20 million that had been granted the project, so they were well within the budget limitations by a very wide margin.

Soon, the glorious Summer weather would turn to Fall, then Winter. August was nearly gone and September was just a couple of weeks away.

Considering all that had happened so far, they were all now on edge and constantly on the lookout for trouble. Even though the Pentagon had stopped the one major criminal organization in Houston, the big oil companies might still present new and different surprises down the road. In a sense, big oil would see this fusion project as a War they could not afford to lose.

Everyone on the team was now fully aware that the closer they came to success, the more they could expect the opposition to rear its ugly head . . . and it surely would.

Chapter 18: Fusion Test #1.

Days turned into weeks and weeks turned into months as the team worked day after day. Throughout September, they tended to trade off between building the third reactor and working on the gravitational focusing of the plasma. Reactor #3 was virtually finished by the end of September. It would need to be tested at the Sandia labs in Albuquerque, but there wasn't a rush to do that.

The team wasn't anxious to go back to Albuquerque, considering all the trouble they had encountered there in the past. They decided to put that off until a later time. Reactor #3 was just a spare now anyway, since #2 had been returned and it was already tested. Once #3 was finished, they set it aside and put it on a back-burner. After that, they all forged ahead, working together on the main compact reactor unit. It was taking shape by late October and considerably ahead of schedule too.

As time passed Enrique became a regular fixture at the house on Arizona avenue. The romance with Audrey had become a full fledged love affair. He was driving up to Los Alamos almost every week end. Sometimes Audrey would borrow Gus's or David's car and drive to Albuquerque to be with Enrique.

Everyone was a little homesick by mid November. It had turned cold and outdoor recreation had become limited. Jim decided a break would be good for everyone. So, at the end of November he had the Learjet flown out and they all took two weeks off to visit friends and family in DC. Audrey was completely miserable without Enrique, but she needed to visit home too, so she went.

The trials were going on for Dugger and his cronies in Houston. The news media had picked-up on the story and they were following the trials. News of the Fusion project became news of the day. It was no longer top secret. If it was, it was a secret that couldn't be very well kept. Everyone knew about it. The major science magazines were all clamoring for interviews. The team was about to shoot towards super-stardom, even though they really had not yet accomplished anything noteworthy or major. Just the promise of fusion was enough to get people excited. Everyone wanted to read about it and learn how it would work.

Denise and the team talked about it after they returned to Los Alamos. They decided to not do any interviews until they had at least accomplished over-unity on the tests. It was appearing as though the new reactor with the electrogravitic vortex assembly might be ready for testing in February, barring anything unexpected. They really wanted to be able to do the tests without the world watching. If they failed for whatever reason, they didn't want to have to deal with the backlash and the embarrassment of failure. Failure was highly likely on the first tries anyway. It's a thing called 'Murphy's Law': "If anything can possibly go wrong, it usually will." Especially on the first attempt.

That's the learning process. Try and fail: solve the problems and try again and again until success occurs. That's the routine every inventor goes through.

Denise was totally diplomatic about the interviews. "No way! Not yet! Not until we are close to success. It was supposed to be a top secret project. She wanted to try to keep it as top secret as possible. Fame would need to wait at least until they had done something worth being famous for! So far, they hadn't done that. The rest of the team agreed with her, including Jim.

When they were at Los Alamos they were pretty well protected from the media. The security around the labs was pretty tight. People weren't allowed to nose around. Everyone there had to have class-a security clearances. It would have become insane if not for that.

The Christmas holidays came and went. Finally in February, the reactor was ready for the first test. They were all nervous about it. They took the completed reactor to a special testing pad that was several hundred yards from the main lab. It was actually in an empty field. The pad had originally been designed as a rocket-launch test-pad but it would suit their purposes and it was far enough away that should an accident occur, it hopefully wouldn't cause collateral damage. There were three, 10 inch thick concrete walls and a concrete pad on which to set the reactor. Of course there were wires and tubes running everywhere. The crew would be nearly 75 yards away at a control panel. It was very similar to a rocket launch. They jokingly prayed that it **wouldn't** be a rocket launch!

They decided to do the test on a Friday afternoon and devoted the entire previous week to setting everything up, then checking, double checking and triple checking everything all week long.

Friday at 2:00 PM, they were all in position at the controls. Gus had rigged a series of dials and gauges that would help inform them of the test results. Actually, everything would happen pretty much automatically and instantly, once ignition had been initiated. It would either work or it wouldn't.

At exactly 2:30 PM Denise said a silent prayer as a count down proceeded. 5,4,3,2,1, FIRE! She pushed the big red button that fired the lasers to commence what they were calling the ignition burn. The lasers would reach 500 trillion Watts instantly and fusing would occur a split millisecond later. It would all be over before that entire second had elapsed.

For less than a second, it appeared as though sustained fusion had occurred. The fusing process had been contained, if only for a fraction of a second. Everything appeared to be as normal as they had hoped for . . . so far. It was a tense uncertain moment.

The objective of this first test was not to hold an indefinite sustain on the reaction. The objective was to simply reach a surplus point called over-unity where more energy had been produced than it took to produce it. They were only hoping to achieve a small surplus.

If they could simply create just one watt of excess energy over and above the amount of energy needed to produce it, they would consider this particular test to be a major breakthrough success, (for them at least). It wasn't a major science breakthrough. This had already been done before at various test laboratories around the World.

In a sense they were following prescribed formula, except they were employing Denise and the team's new electrogravitic assembly, to help increase the pressures on the fuels during ignition. The intention was to fuse the near microscopic, mixture of deuterium and tritium in a more efficient way than had ever been tried before. They hoped the tiny amount of fuel would fuse and produce an excess of heat energy. If all went well, it would occur within a fraction of a second and the electrogravitic assembly would prove or disprove its viability.

That would be it! The test would be over. They would let everything cool, and finally they would inspect the reactor and the vortex assembly to see how well everything had held up under the pressures and the heat.

The test was over with a snap, as it began.

Nothing blew sky high! Nothing exploded. Everything appeared to be normal. The reactor sat in silence. A violet glow surrounded the electrogravitic assembly housing. That was normal and expected. That second was pretty much anti-climatic. There was absolute silence for a very log time as they all checked the gauges and sweated blood. Everyone had been holding their breath. They were all in a state of reserved judgment, checking their dials and meters.

Then they all went crazy, cheering, laughing and slapping high-fives. As far as they could tell at this point, the test had been 100% successful. The gauges and dials were showing a 10 X surplus of power.

They would let the reactor cool off over night, or however long it might take, then sometime tomorrow they would inspect for damages. Of course they were hoping that the vortex assembly had completely done its intended job. That was hoping for a lot! It would tell them whether or not the Starshine Project was on the correct path. They fully expected to see at least some damage. If there was damage, they would analyze it and find ways to avert it for the next test.

If there was no serious damage, they would use increased amounts of fuel and do longer and longer tests over the next few weeks and months with the objective of one day maintaining a sustained reaction for minutes and possibly even hours. They were still a long way from that dream but today's test was apparently successful.

Jim fired an e-mail off to General Frank Wilson:

Starshine successful! Surplus energy on test #1!
Fusion sustained for 1/1000th of one second.
10 X Surplus energy achieved.

Jim broke out a bottle of Dom Perignon on the spot and they celebrated with plastic glasses. One major milestone had been reached. There would be other roadblocks ahead, but this was the potential beginning of a brand-new World.

Chapter 19: Sustained Surplus Fusion Energy!

The following day they all returned to the lab to do a thorough inspection of the reactor and everything inside it. They would tear it down and rebuild it, while searching for flaws and problem areas. If everything looked perfect, they would all breathe a sigh of relief. They would then reassemble it, add a tiny bit more fuel and do a (fraction of a second) longer test on the next run.

If there appeared to be any problems or weak places, they would deal with each one individually and try different things to hopefully solve the problems. It would be a trial and error process from here on, until absolute perfection might be achieved.

As they carefully disassembled the reactor one-piece-at-a-time, they were amazed at how well the vortex assembly had performed. No overheating of the crucial and sensitive parts had occurred. They were overjoyed about that. They found a few singed areas that might need attention. The singed areas might possibly burn and cause problems in a longer test. Gus and David went immediately into MacGyver mode. Within three hours they had contrived safeguards to protect those singed areas and pull them out of harms way. Everyone was convinced that the changes would only lead to better performances in the next test. It took most of the following week to reassemble the reactor, but a second, longer test, was scheduled for the following Friday. It took all day Wednesday and Thursday just to set it up and be certain everything was a go. The test was set for 11:00 AM on Friday morning.

On Friday at 10:30 AM everyone was seated at the control panel where they could immediately read the gauges after the test.

Everyone said a silent prayer that it would be a successful sustained fusion reaction and NOT a rocket launch, (which had become a standing joke).

At exactly 10:59:55, they did an official countdown and as the clock struck exactly 11:00 Am on Friday March 17, 2017, Denise hit the big red 'Start' button . . .

A millisecond later, all that could be heard was click or a snap. It was over! Ignition and fusion! Then, they all observed the power input and output gauges. The output had reached surplus and had gone well beyond their expectations. They had achieved 150 X surplus! The excess energy was 150 times the input. The test was still limited to only a fraction of a second but the surplus numbers were what counted.

Everyone held their breath for a long second as they tallied what the gauges were telling them.   They were afraid to dream even for a second about what it actually meant . . .

SUCCESS! 100% SUCCESS! They had done it. It was a successful sustained reaction which had produced 150 times more energy than input energy used! Now . . . they had done something really special! This had not been done before as far as anyone knew. It was unprecedented.

Like the caveman who once rolled a wheel down a hill and made a wagon, they had just taken the next forward leap in the evolution of all mankind.

The champagne corked popped and went flying! The cheers went up . . . even louder this time! They had actually done it! It seemed unbelievable, but they had done it well under budget by over 17 million dollars. They had done it in exactly one year . . . not four. Yes, they would become very famous. That had just become an obvious fact.

It wasn't over. They weren't finished yet. Actually it was only the beginning. The true objective would be to maintain a sustained reaction for indefinite periods . . . not just a fraction of a second. Just the same, it seemed totally obvious that if the unit could hold up for for a second, it could surely go for much longer periods. This was truly a milestone to shout about and shout they did!

Little did they know there was a plan already underway to thwart their short lived success . . .

763 miles to the southeast, in a dark and vacant warehouse near a railroad track in a run-down area of Houston, Texas; seven people hovered at computer terminals and laptops. The spacious warehouse was otherwise completely empty. Only one light glared from a distant naked light bulb on wall. Seven computers were lined against one short wall. A bluish and eerie tint illuminated the bleary-eyed odd assortment of humanity at the

terminal keyboards. There were five guys and two girls. The warehouse was so large their voices echoed when they spoke. Most of the time they were all silently pecking away at their keyboards without speaking.

They were all younger than 22 years old. Both girls appeared to be in their late teens and their style of the moment was Goth. The two girls were wearing black t-shirts and black jeans with black shoes. They wore their haired bobbed and jet black with much too heavy mascara around the eyes and down their cheeks. It gave their eyes a dead, empty hollow and weird look. One girl had scarlet streaks in her jet black hair and they both wore scarlet lipstick.

A poster for a band called 'Death Valley Scars' was tacked to the wall.

Not all the young guys were Goth. One of the young boys in the group had natural flaming red hair and he wore it long and over the shoulders. He was probably 18 or 19 and skinny as a rail.

They were a strange looking crew. They called themselves Hashbash. This group Hashbash, a hacker group; was being paid extremely well to do what they loved to do. They were all hackers and each was a master at it. The IQ levels in the room may have been off the charts.

Hashbash has no idea who they are working for and they don't care. A courier who goes by the name of Phony Tony knows everything about this group of computer fanatics. He had found them on the Internet, down in Usenet . . . (the darker underbelly of the World wide Web).

Hashbash has a world-wide reputation as one of the most elite hacker groups on the planet. In spite of their young ages they are great at what they do. They have all been hacking for the fun of it since before they were teenagers. Hacking is their play, it's their fun, their recreation, their reason for being and it is also now, . . . their very lucrative vocation. They have graduated into the big leagues with this particular mystery job.

Phony Tony the courier had made contact and brought a briefcase full of money. They were being hired to do a specific hacking job for an unknown and obviously extremely wealthy mystery client. They were not to ask or investigate the client's identity. They were not to investigate the courier's identity. That was as a big part of the deal. They would be paid fantastically well to quietly do a job, then just as quietly, disappear.

Phony Tony had worn a disguise both times they had seen him; once when he solicited them for the job and once again when they were paid a deposit to begin working. They would see him just one more time, as they were to be paid a much larger amount at the successful completion of the job.

Hashbash had no clue what Phony Tony actually looked like or what his real identity might be. He always wore a disguise and it was a non-issue. They all used fake names too. Secrecy and hacking go together. All they cared about was the money. This odd assortment of hackers called Hashbash had been friends throughout all their school years.

The reason Hashbash had become friends was because of their mutual deep-rooted need to hack into computer systems. That was how they had met and it was the thing that bonded them together. Hacking was now their vocation and they were determined to do the job perfectly. They had recently graduated into the big leagues with this particular mystery job.

This particular hacking job was actually fairly simple. It would have been near impossible, but someone had furnished both the password and the uniform resource location code or path to the computer to be hacked. The fact that the computer to be hacked was behind one of the best firewalls on the planet was a non-issue; because they already had the password. It should be an easy, in-out-and-done, job. Very easy money!

The only request was to download and save a gigabyte of files, then delete everything in all the folders and backups at the specific FTP address. How difficult could it be?

It never occurred to these kids in Hashbash that what they were doing was a very serious federal offense. This job could possibly get them locked in a federal prison for years if they were to get caught. They were hackers! That's part of the fun of being a hacker. It's all about beating the machine and beating the system. It's a game. Good hackers do not get caught!

Chapter 20: Cyber Attack!

Back in Los Alamos, the process had become a routine. They would allow the reactor to cool during the week-end and on Monday they would disassemble and inspect it thoroughly again. They would look for flaws and problem areas then try to avert any potential problems before damage could occur, then they would test again and again, increasing small amounts of fuel each time and extending the fusion timing lengths.

It was Friday evening. Jim was so thrilled with the day's test results he had a huge feast catered at poolside. They all partied as they had never partied before. They simply could not contain their exuberance. What they had just done on this day was huge. It meant that the World would never be the same . . . because of them! They were now in a league with Oppenheimer, Feynman, Groves and all those great people who so long ago had been a part of the World changing, Manhattan Project. Project Starshine has just gained equal status with the Manhattan Project! That was something to celebrate.

Enrique was there with Audrey. He was now considered one of the gang. They had finally let him in on the secret of what they were doing. He was shocked to realize that Audrey wasn't the department store clerk she had claimed to be, she was now, not just a physicist but a soon to be, world renowned physicist. He shook his head in disbelief, but this particular party at poolside was a testament. He was happy and proud for everyone on the team. His affair with Audrey was looking very serious. Marriage had been mentioned more than once. The two were apparently head-over-heals in love.

The party lasted well into the night and no one emerged totally sober. By 3:00 AM Saturday morning, they were all partied out.

On Saturday after the celebrations, Denise went over to the labs after lunch, with a mild hangover, to do her weekly 'think tank', as she had started calling it. She was simply a workaholic. She was passionate and addicted to her work. It was all she ever thought about. Poor Jim felt somewhat abandoned but he too enjoyed a little 'me' time for himself. The couple was very much in love, but their love simply wasn't the clinging, *need-you-every-second*, type of love that Audrey and Enrique seemed to be enjoying.

On this particular Saturday, Denise went inside the lab to study the diagrams and equations on the computer. The reactor was on the pad out in the field about 100 yards away. She would leave it be until Monday.

As she fired the computer to access the folders, a bright red *** WARNING!!! *** screen greeted her in a very ominous way. There had been a serious security breach. As she attempted to pull up the reactor folders on the computer system her heart fell into her shoes. The folders were completely gone. Deleted! Something was seriously amiss.

All their work was gone! The latest successful diagrams for the reactor and the vortex assembly were gone. All the equations that sometimes took weeks to solve . . . were gone! Everything was gone! She went to the backups. They were gone as well! Deleted! What could it mean?

It was obvious. It was a cyber-attack! Someone, probably a hacker had been into the computers and all the highly classified files. They had probably downloaded everything to their own computers then deleted everything from the Los Alamos computers; even the back-up files! It was immediately obvious. They had been hacked. She had older versions of some of the diagrams and equations on flashdisks but it wasn't enough to rebuild a perfected electrogravitic assembly and reactor from scratch.

She was totally perplexed. A computer might sometimes mysteriously delete a file or two but not files like these. Only five people knew the passwords to these files. The system had a firewall like few other computers on the planet. This was Los Alamos! It just could not happen. She was fairly certain that the files could not be hacked from the outside. Not without the passwords. It was a virtual impossibility.

She called Jim on the cell and explained what she was seeing. It could mean a lot of things, and each was a terrible and horrific scenario. Whomever had those files could use them to build their own reactor. Not only that. It was worse than that. There wasn't any way on earth that anyone on the team would be able to remember each and every solution to each and every equation. They couldn't easily replicate the reactor without those files. They couldn't even resume work on the current reactor sitting on the test pad.

They referred to those files on a daily basis as they worked on the reactor. Those files were necessary to their every day progress and without them they would be set back by many months.

Jim tried to remain calm and cool, but the more Denise cried on the phone, the more upset he became. He thought at first that they would recover easily, since they were the inventors. That just wasn't so. There was simply too much to remember and too many intricate and complicated details within the equations and diagrams. As it stood, they were currently at square-one. They were essentially starting over!

Jim borrowed Gus's car keys without asking and ran out of the house. He was at the lab in 5 minutes. Denise was sitting at the computer terminal with her head on the desk, sobbing almost uncontrollably. Jim quietly walked in and began massaging her neck from behind as he tried to think of every positive phrase he could think of, to make her feel better. She stood and they hugged for a very long minute as he tried to console her and she struggled to compose herself.

She spoke first, "We do have access to the working reactor on the pad. It could eventually be reverse engineered. It would be easier to reverse engineer the thing, than it would be to rebuild another one from scratch without the diagrams and solved equations.

They needed all their previous years of work. Some of those folders included equations and diagrams that dated back to when they had formed the team at MIT. The cyber-attack actually meant total devastation to the entire project, no matter how optimistic she tried to think." Then she broke into sobs again. Jim held her in his arms until she had calmed a little more.

Of course as soon as she had settled a little, Jim was on the phone with General Frank Wilson at the Pentagon. He caught him relaxing at home, "Jimbo! It's good to hear from you. How's it going? Congratulations by the way. I've sent a bottle of Champagne. It should be arriving at any moment. We are all so proud of you guys. I even called President Clayton personally and bragged that you were able to bring-in the reactor for well under the proposed budget. She was thrilled. What's up?"

Jim actually stuttered as he began, "y-you won't believe it, Frank! We've been hacked! It is triple serious. A hacker broke into the Los Alamos computer system and not only did they steal all the diagrams, the equations and everything necessary to build a working Compact Fusion reactor; they also deleted everything.

It's the deletions that worry us most. Denise and the team are stymied without those files. It's all too detailed. They will never be able to replicate the success without rebuilding it all again from scratch. It's an absolute nightmare, Frank. We've just discovered it minutes ago. We're totally clueless and completely panicked.

We don't know where to turn or what to do about it. It could be a hacker . . . anywhere in the world. It could be some geek in China, North Korea or Iran for all we know. We hope you may have a way to help us sort it out. We're down for the count at this particular moment in time. We don't know where to turn, or where to begin looking."

General Wilson was known for his violent temper. He virtually exploded over the phone. He was insane with anger. The profanities rang in Jim's ear for nearly a minute before the General calmed enough to begin apologizing. "Jim, Jim, Jim . . . I'm so sorry. My anger isn't directed at you. I'm just blowing off steam. We will get them and we will hang them by their thumbs. It's that rotten-to-the-core Houston bunch again. I think we can make that bet and win it. They're an unscrupulous gang of barbaric thieves. We will get them. Don't you worry.

Here's what we need to do. I'll send a team of my very best computer people out there to Los Alamos. They will track everything that can be tracked, from your end. We'll get a clue that way, then we'll follow it. It will lead to someone somewhere, who will never be able to hide from our people. We will get them and they will know what pain is when we're finished with them.

There will be a team of specialists out there within a matter of hours. We keep teams of people ready to travel, just for situations like this. Our best and brightest will be in the air within the hour.

Secure those computers! Do not let anyone touch them. Tell the director of the lab that, by order of the Pentagon; they are not to touch a thing. Wait for our people to get there. Our people are all top of the line. They're the best in the World at what they do. They know every computer trick in the book. They will follow the trail to the source of the trouble. You call the director of the lab right now and put him on alert. I'll call him as well, just as soon as I get the our hacker team in the air.

Spell out the word T-R-E-A-S-O-N very slowly to the lab director. Tell him I will be in touch within the hour. Our team will be there in under six hours. The lab director is to touch nothing! Don't even send one e-mail and do not logon to the Web until we get there. Do you follow me?

Jim, don't worry! We will catch them. That's a promise. I'm hanging up. The wife and I were headed out to dinner. That's off, and she's gonna be mad. It looks like it'll be a working Saturday for this old General. This is the point where I need to step in and earn my paycheck. Stand-by. Our team of computer experts will be there in six hours or less. See ya Jim. Stay cool, buddy! Gotta go! We will catch them!" He hung up the phone.

Jim sat at the desk with his head buried in his hands for all of ten seconds, then he was on the phone to the lab director.

Within a minute he felt warm hands caressing his neck and shoulders from behind. Denise was doing what Denise was so especially good at. She had put her own emotions in a box and placed them out of sight. She had never seen Jim this upset since they had met. It was her turn to console him. "We'll sort it out hon. It's just a bump in the road. We will emerge victorious. Don't worry." There were huge tears in her eyes and flowing down her cheeks as he looked up at her. She was as devastated as he was, but crying wouldn't fix a thing. They both knew that. They decided to go back to the house.

As they arrived back at the house with both cars, the team was out by the pool. Up to this point, none of them knew what was going on. The minute they looked up and saw the grief on Jim and Denise's face, they could tell instantly that something really horrible had happened.

Jim called them all around the table and everyone grabbed a beer and a chair. Enrique and Audrey were there, along with Gus, David and a beautiful young blonde that Gus had invited over for a swim. She had been over to the house, spent the night with Gus and had become an occasional fixture at the pool. She was extremely easy on the eyes but according to Gus, it wasn't serious. Gus was just being Gus. He liked young and beautiful, stupid women. Her name was Brittney McGill and she was by no means on an equal intellectual level with any of the crew.

Jim and Denise took turns explaining the situation. Every face at the table went from happy to complete devastation as the story unfolded. Even the blonde had a look on her face that made her appear to be a deer in the headlights. She looked terrified and she actually had no clue what they were talking about.

Gus began wringing his hands and walking in circles. It was his thing. He did that every time he was stressed. Audrey burst into tears. She was sobbing uncontrollably.

Enrique did his best to console her. Enrique didn't seem sure of what it all meant but he was bright enough to sense how devastatingly bad it was. He looked a little like a rabbit caught in a trap too.

No one noticed but Jim, he had trained himself to read people by watching the eyes. The eyes never lie. Something wasn't right about that look in Enrique's eyes. Was he masking something other than automatic concern for Audrey?

Gloom and doom became the phrase of the day. The more they thought, the worse the situation began to seem. All their complex equation results had evaporated into thin air. Some of those equations had taken weeks to solve. The recent diagrams were just as important. They had been slowly built-up over the years by trial and error and the trial and error they depended on almost exclusively was the ever evolving equation solutions.

At 9:12 PM Jim's cell phone rang. The team of computer specialists were already at the local airport. They needed a ride over to the lab. The whole team went. They took all their cars, partially because of nervous energy that needed to be expelled some way, and partially because there might be an outside chance they could help rectify the situation in some small way, just by being there.

Enrique received a sudden desperate phone call from his mother in Albuquerque. She was deathly ill and wanted him to come home immediately. He shook hands with everyone; hugged Audrey for a very long time; wiped the tears from her eyes and spoke softly to comfort her as she walked him to his bright yellow Jag convertible. A few minutes later he pulled out of the driveway and was gone.

The 'hacker-team' of computer specialists was made up of three very young looking guys. One was tall and lean. He sported a Mohawk haircut and a full beard that hung almost to his chest. He dressed in baggy jeans that hung far below the hips almost to the knees. His t-shirt advertised some band no one had ever heard of. He was one very strange looking dude. His name was Mike. He didn't offer a last name.

The second member of the team was an obese kid with horn rimmed glasses, curly black hair and a mess of pimples on his face. He too said his name was Mike, but since there seemed to be a run on that name, . . . he would answer to Mikey.

The team leader looked almost like a normal human being except for the bushy mass of hair that hung off his shoulders in kinks and tangles. He too was named Mike but he would answer to the name of Michael.

These guys looked a lot like a cartoon walking! How could they possibly be the cracker-jack team of hacker specialists that the General had talked about.

The entire Starshine crew looked at each other in abject bewilderment. They were all judging by what they saw in front of them. Hacking is a special talent, requiring a very special set of skills. Hackers just don't have to conform to anyone's rules. The oddball team of guys named Mike all carried a small overnight bag. That's all! It was obvious they didn't plan to stay long.

Since there were three cars there at the airport, each Mike crawled into a different car and everybody headed for the lab. Soon the lab was full of people, the entire Starshine team and Team Mike; (the computer specialists), all crowded around the computer terminal.

Team Mike asked the Starshine Team to all to please wait outside the lab, while they worked. Within seconds the computer terminal lit up and the three Mikes talked softly among themselves

The Starshine Team gave them their space and walked down the hallway to a small vending room that contained snack machines. There was considerable snickering and giggling going on with Team Starshine.

Team Mike looked like a crew of wayward runaway kids. They certainly didn't look like computer wizards from deep within the bowels of the Pentagon, but they all carried security credentials that allowed entry to the secure area. The security guards hadn't said a word. They simply nodded with approval and allowed them in. Everyone on the Starshine Team became a little worried that they were being duped, but they held their tongues and waited outside. Team Mike went to work, doing what they were there to do.

About an hour later, Michael (with the bushy hair) emerged, smoking a joint. He grinned and offered a toke to everyone there. He was actually a pretty neat guy in spite of his arrogance and appearance. Gus and David both took sociable tokes off the joint, mainly to make Michael feel more comfortable. They didn't want to give the impression that the Starshine team was snooty or snobbish.

Michael (with the hair) began to talk. "Well we were able to detect a path. Whoever broke into the system had solid inside information. Your password is as good as passwords can be. It is encrypted with SHA 512 encryption. Passwords can't get much better than that. It's virtually unbreakable. They knew the password or they wouldn't have been able to access the files. We know this because we've also checked the time tables. They were inside your computers on their first try, within 5 minutes from the time they had begun logging in. No password can be cracked that fast without prior inside knowledge.

So, the bad news first: there is most likely a mole among you. That's our first suspicion. One of you 5 is guilty!" He pointed a finger at each of the five Starshine team members individually; looking deep into each of their eyes as he did it. For a second or two he appeared to be a trained New York City detective, grilling a perp. He was looking for the truth by reading the faces.

"Who ever that person is; also has contacts in Houston. We traced the files. They went from your computer through nineteen different nodes around the World, but they ended up on a computer in Houston Texas. We now have the exact street address of that computer and we've notified the FBI in Houston. There is a raid taking place on that computer as we speak.

It is really good that you discovered this breach when you did. The breach occurred at 1:47 PM and someone named Denise discovered it at 2:33 PM. The perps surely weren't expecting that to happen! She almost caught them red handed, (not that there's much she could have done to stop them).

Anyway, the fact that she discovered the breach so quickly and we've been called in as quickly as we have, essentially means they are toast. We've caught them pulling up their pants, so to speak. They are probably wearing metal bracelets right about now and they're wondering how we've caught them so amazingly fast. They probably thought no one would have checked the computers over the week end. That would have given them time to sweep their trail. We've got them. No problem there.

Who is Denise? Which one of you is Denise?"

Denise raised a hand and mumbled, "I am."

Denise, you saved the day by reporting this as quickly as you did! Since she reported the breach so quickly, it makes her above suspicion. The true mole would not have reported it that quickly. So, she is off the hook of suspicion. That brings the list of suspects down to four.

But . . . YOU GUYS have a serious problem here! The way we see it, someone here, either knowingly or unknowingly, has passed or given up the password to someone very bad, in Houston, Texas. The hacker crew in Houston is apparently connected to the big oil corporations that base their operations out of there. We think it is a secret consortium of several companies, but until now we haven't been able to name names. We are looking to break them up. This may be the day we do it. We've been hoping to catch them red-handed for a long time. This very well may be that day!

Just the same, you people need to start looking at each other right now. There are only four potential suspects. Which one is it?

We'll leave you to settle that. We're done here. Our report to the General will be worded about the same way I've just said it to you. I'd suggest that you'll be wise to figure out who your mole is before the General gets our report. He can be a real nasty SOB. I know him too well. Do you know what I'm saying?

Can we get a lift to the Airport?" The kid smiled like a Chessy eating a canary; tossed the marijuana roach he was smoking to the floor and stepped on it with his foot.

The other two Mikes emerged with their handbags a few seconds later. They hadn't spoken much more than their names since they arrived.

The Starshine team was immediately perplexed. They knew with absolute certainty that none of them would be a mole for any kind of money! They had all poured their hearts and souls into this project since day-one.

They didn't know what to think . . . unless it was Enrique! He was the obvious choice and Audrey realized it first. She gasped, then choked. Then began saying "My God! My God! My God! How could I be so stupid. I believed him. I believed every word.

He was so sincere. How could he do this? How could he? He must have snooped my purse to steal the passwords."

The sobs began soon after, and they weren't going to stop anytime soon. This was going to hurt her and it was going to hurt her deeply. Not only had he fooled her emotionally all this time. He was a plant from the very beginning. He had hit on her at the night-club in Albuquerque. He had known at the time who she was and what he wanted to do. It was vicious and vile. He had been feigning love and had been using her all along to discover the password.

Denise immediately went to Audrey's side. "Come on girl. Let's go, hon. Well go somewhere and talk. I'm with you girl. You couldn't have known! None of us suspected a thing. You aren't the only one who was fooled. He fooled all of us! We'll go somewhere quiet and talk it out. I'll get you through this kid. I'm with you! We're all with you and no one is going to blame you. We understand. We know you're innocent."

Denise looked at the rest of the team with a sadness that can't be expressed in words. She also offered a warning look to everyone there. It told them all to stay back and stay cool. Then she looked directly at Jim, "Jim, let us take the Mustang. You guys get Team Mike back to the airport. I'll handle this. Audrey and I just need to talk it through. We'll be okay. Give us some space and some time, okay? "

Jim tossed her the keys with a wink. He knew she was being more than the team leader. She was also being the best friend that she could be, and Audrey needed that particular friendship at this moment in time. The two girls were extremely close friends, first and foremost.

Team Mike was escorted to the plane. David and Jim went home to the pool and were sopping up beer and practically crying in it. What a nightmare it had been. Their conversation was all over the place. They worried for the project, but the conversation turned brighter as they realized that the files would soon be returned and the hackers were probably already in jail.

Then they began to discuss Enrique and his connection to the whole thing. They had all been totally taken-in and completely fooled by him. He was definitely and obviously very smooth and without any soul what-so-ever. He would be found, and he would do time. What he had done was a federal offense. It qualified as treason. He could get life in a federal prison. He surely knew it.

He would probably be on the run for a long time. He had obviously been paid a considerable amount of money by someone big, in Houston.

As they all weighed the evidence one thing seemed to be in conflict. Enrique simply had to have some sort of feelings for Audrey. No one earth could ever be so cold as that. Could they?

That was the question of the day . . .

Chapter 21: Recovering.

By Monday Audrey was looking a little better. Her eyes were hollow and her face was drawn. She looked terrible, but she occasionally forced a smile as the team constantly helped her realize that it could have happened to any one of them. They assured her that it wasn't her fault; and if any of them had even suspected Enrique of anything so heinous, they would have beat him to a pulp. It made her feel a little better to know the team was 100% behind her all the way, but that didn't stop the pain she felt for Enrique. She had fallen deep and hard for the guy. He had seemed totally sincere the entire time. How could anyone do such as that? It was beyond all of them, but that didn't stop the hurt. Audrey was crushed.

She had e-mailed Enrique a scathing e-mail, admonishing him for what everyone was sure he had done. He was the only one who could have done it! She called him every foul name in her vocabulary.

She had received a haunting reply from Enrique. It simply stated: "I know you won't believe it Audrey, but I didn't do it. Sure! It looks like I did, but apparently that was the plan. I've been framed. I'm closing this e-mail account. You won't find me until you find the real traitor. I DID NOT DO IT! I love and miss you." ~ Enrique.

Audrey deleted the message in anger as she read it, but she did tell Jim that Enrique was pleading innocent. Of course he would plead innocent!

Jim hugged her neck, "If you get another email like that, save it. Check the headers. Save the headers especially. We'll send everything to the General. His hacker team may be able to track the headers of the message back to Enrique. If he's innocent, we need to know it, and we need to still be looking for our mole. He's assumed to be guilty because there's apparently no one else it could be.

A light bulb went off in Jim's brain. The cute, dumb acting blonde that Gus had been with several times, Brittney McGill. She had escaped scrutiny, simply because she seemed too dumb to be involved in a plot such as this. What if . . .? He let it rest in his mind and didn't say anything to anyone but he allowed for the fact that she could be smarter than she had let on.

Good news came from Washington around noon. General Wilson called and told them to check the computer terminal while he waited on the phone.

The files had all magically reappeared exactly where they belonged. He informed the team that a ring of computer Hackers had been caught red-handed. There were seven of them, all kids, five guys and two girls. It was a hacker group called 'Hashbash'. They were all in jail. It was very unlikely that any of them would be released anytime soon. They were being charged with treason against the United States of America. So far, they had not divulged who exactly they worked for. It seemed that even the leader didn't know actual names and/or specific corporations. An unknown courier in disguises brought messages from above and faithfully made promised payments, otherwise . . . they didn't appear to know who exactly they were working for.

Agents were on the trail of Enrique Smith. That name was phony, by the way. His real name was Manuel Ramos Rodriquez. He was an illegal alien from Cuba and it was likely that he had returned there. Enrique would know that Cuba would not extradite him back to the US for his crimes. He would probably remain safe as long as he stayed in Cuba. But; the FBI was like a bulldog. They would never forget who he was and what he had done. If ever they had a chance they would hang him.

The Starshine team needed to get back to work on the project. The long week end was over. They were all totally exhausted from the ordeal. Their lives had been roller-coaster crazy since Friday. They had gone from the extreme, outrageous high of being equals with the Manhattan Project on Friday, to the extreme lows of being robbed of everything on Saturday and then back to the highs of having all the files returned today, which was Monday. Emotionally, everyone on the team was a total train-wreck in progress.

Jim called a meeting of the minds. He felt pretty sure that they all needed a mental break. They took a poll and he had called it right on. Denise was the only one who was actually ready to return to work on the project and she was simply trying to be tough. She was an emotional wreck as well.

After a considerable amount of cajoling and ribbing, she gave in to the crew. They decided to take Monday off. They went to the Blue Window restaurant again for prime rib and a fresh bottle of Moet & Chandon Dom Perignon Charles & Diana 1961 Champagne; compliments of General Frank Wilson. The Champagne had arrived late, but in reality the timing was perfect.

The team needed any reason they could find to smile. By the end of the day their moods had lifted considerably. The waitress and waiter who served them were champagne connoisseurs. Both claimed that the champagne they were drinking was list priced at more than $4,000 per bottle. General Wilson certainly wasn't cheap when it came to champagne.

Audrey drank twice what she should have, but she vowed to not let the creep Enrique get to her. He wasn't worth crying over for another second. She was a big girl and she would get over it. She was beginning to look and sound like her old adorable self again. She would survive.

On Tuesday they all returned to the lab. They were bright eyed, bushy tailed and ready to go. They disassembled the reactor unit once again and inspected it once again. To their surprise there was no apparent heat damage. It appeared the gravitic assembly was working exactly as planned. They would be free to try a more extended test as soon as it was reassembled and by Friday it was on the pad and ready.

Each new test pushed the envelope. Within a few weeks, they had pushed the fusion time to nearly 5 seconds. They had produced approximately 5000 times the energy input. This meant they were far, far ahead of their own expectations and they had completely blown past all the competition by years and miles. To imagine they had surpassed 5000 times input meant major changes in the testing.

They setup a 100 megawatt turbine generator and funneled the heat to create steam that ran the turbine to produce electricity. Finally they hooked into to the grid and would begin producing free electricity to the small town of Los Alamos for a short while on their next test. That test was setup and ready to go. It was a simple matter of hitting the switch.

As they were setting all this up, which took almost a month, the media became aware of what they were doing. Los Alamos was immediately swamped with news media from practically everywhere in the World. Every major news network begged to be privy to the experiment. It was unavoidable. They were about to go public. No one was anxious to do that. It still felt like the timing was too soon. They had wished for more time to experiment in seclusion, but it wasn't to be. They were boxed in. This was the next logical and most practical step. Schrodinger's cat was out of the box now, very much alive and would be doing quite well forever. The future of fusion power was pretty much assured at this point in time.

Jim flew General Frank Wilson in on his Learjet. It cost him a bundle but he didn't care. He did it simply because of their friendship. The Learjet was far more comfortable than the military jets the General had available to him and quite honestly, it was easier for everyone involved.

General Wilson flew in on a Wednesday just prior to the big Friday test. Jim had the extra bedroom and General Wilson stayed with them at the house on Arizona avenue, rather than the Officer's Quarters at the base. They would all get a chance to hang with him at the pool, and get to know the guy who had made such a great difference in all their lives. Up until this point those on the team had only seen him briefly a couple of times.

The General was a mountain of a man. On arrival, he was wearing his Army dress uniform. It held numerous medals and badges all down the front and he was a picture of authority. The three stars on his epaulets glistened in the sunlight as he stepped off the Learjet at the airport with a million dollar smile.

He was in his early 50s with graying hair and easily stood at around 6'2", weighing approximately 200 lbs. His blue eyes were like the lasers they used in the reactor. It seemed that he could bore holes with them and his pearly grin which was almost too perfect. The teeth almost certainly had to be false. But, they looked perfect. He always offered a warm and friendly smile, until he had been crossed, then it was easy to imagine how those same teeth might begin to resemble those of a Great white Shark. General Wilson was a formidable appearing presence for sure.

This particular General was one of the most powerful people on earth as head of the DOD Research and Development. It showed, in his personality, his demeanor and his physical presence.

Within a hour of his arrival the team was all gathered around a patio table at the pool, drinking margaritas and other alcoholic beverages of choice, chatting with the General. Denise and

Audrey grilled steaks at the propane grill, within earshot. The mood was festive. The General was a fun person to be around. He had changed into an expensive looking bathing suit and a tank top with flip-flops, as he sipped a frozen margarita. If you didn't know who he was as he sat there by the pool you would just think he was an average, successful middle-aged man.

Gus's eye candy, Brittney McGill sat on his lap in a string bikini that left very little to the imagination. Everyone was sitting and listening attentively. The General had their attention, so he launched into an informal meeting of the minds.

"Here's the way it is all stacking up," he said as he took a long pull on the margarita straw. "It appears that we've reached critical mass with this wonderful little project. I guess we are all about to share some limelight whether we like the idea or not." His eyes seemed to rest on Britney more than anyone else in the gathering. She was a looker!

"Public relations are what they are! They're a pain in the neck. Sometimes we hate catering and pleasing the press and the people, but after all, when push comes to shove, they're the folks footing the bill here. They're paying for all this. PR becomes a very important part of the job.

We'll orchestrate and rehearse a press conference for this thing on Friday. We'll tell them that this is a top secret government project. Well do the one short press conference, and that's all.

We'll wait until the test is successful, then we'll all drive to the security gate where the press will be allowed to gather and setup their equipment.

We'll have an MC introduce me. I'll say a few lovely words about you guys, then I'll introduce Jim. Jim can say what he wishes to say, then push the ball to Denise and the team.

Jim, you and I will evade all the technical questions they will be tossing at us. We'll simply tell them to wait until the team is introduced. Denise and the team should take turns answering all they can stomach. There aren't a lot of rules about that. The science mags and reviews will publish everything you are doing in minute detail anyway. You already know how that works.

You'll publish a paper explaining everything; then all the negative nay-sayers and pessimists wannabes will tear it apart, claiming why it will never fly. They'll snatch it out of the air and stomp on it to prove their point, if they get a chance. The Wright brothers went through a lot of that, and Edison too. Tesla was completely destroyed by it or we might have been enjoying fusion technology a long time ago. I understand that he was the initial inspiration to T. T. Brown. No telling where he might have taken the World if they would have let him. He was shot down, then buried and forgotten by bad press and unscrupulous entrepreneurs.

There is a certain element of the science community who actually do little else than sit around claiming that everything is impossible. It's the only way they can gain any recognition at all. They know they can ride pretty high on your fame, as they do their best to discredit whatever you've done. People will listen to them for awhile, but not for long. Those types usually die alone and completely forgotten.

Denise, you guys remember, this is still classified top-secret. That's really all you have to divulge. I'd say that you shouldn't divulge all your secrets. If the questions seem too sensitive, just say those two words: Top Secret!

We want to make a good impression, get it over with and move on. Are we all on the same page? You guys will be getting offers from the television talk shows. You are the biggest news in the World already, but as soon as that test is over on Friday . . . this whole thing will go totally ballistic. You will each become household names. That's a scary proposition. It seems great at first, then it turns into a curse. The next thing you know, you won't be able to go into a public bathroom without an audience following and asking stupid questions while you're trying to do your business. It can get that bad and even worse. Your intimate lives will become front-page news. They'll go to your family and friends. They'll hound them as a way to try to reach you. At some point you will want to become reclusive and just hide from all of it.

So, I'm warning you up front." He was looking directly at Jim and then Denise. "Spread the kudos and accolades all around. You are all a team. Each of you have contributed as much as the other. Give credits where credits are due. Share the fame. There will be more than enough to go around.

I'd say in the beginning that no one should do a talk show alone. Do everything as a group if possible. I don't want to see a break-up, like the Beatles, . . . over egos. If I see any one of you taking too much credit, I'll step in and straighten you out. Egos are the worst part. You are all ambitious and you are all geniuses. Spread the fame and keep it fair!"

Soon the steaks were ready with all the trimmings. The food was delicious. Like a small little family they partied into the night, swimming, drinking and having fun. On Thursday they laid around all day soaking up the sun and relaxing. There would be a few last minute checks to do on Friday morning. The big test would be at 1:00 PM and would last only a few minutes. It was just a test to prove the feasibility of what would soon be coming to the entire World.

They all agreed to dress for the occasion on Friday. The guys would all wear suits and the girls had gone to a fancy Boutique and found great looking business dresses that made them look especially beautiful. They both could have worn coveralls and straw hats; the World was about to fall instantly in love with each of them, either way. Both, Audrey and Denise were Hollywood beautiful, no matter what they wore.

Chapter 22: Going Public.

When Friday morning arrived they were all edgy. It was a big step. The test itself was worry enough. No matter how the test went, they would still be facing the press afterward. That was scary for all of them. David and Gus were turning cowardly about the whole thing. They offered to let the girls handle the press.

Denise bristled over that suggestion. "No way! No how! We're in this together. I plan to pass the buck to you three at my first opportunity! I'm as terrified as all of you. Audrey and I need your support. So . . . put on the big boy pants and man up!"

Everyone laughed. They were all equally terrified.

They had set the test for 1:00 PM. The entire press corps had been alerted to that. They were told that there wouldn't be a lot to see. At best, the lights might blink once or twice. Three different networks had setup a light bar with nearly fifty 100 watt bulbs, just in case they might blink, the cameras would catch it. TV being a visual medium, they were trying to show the exact second the World would change forever. That would be about all they would get for their effort. The whole idea of the test was to provide a smooth transition over to fusion and back in less than a few seconds, just to show that it could be done.

Finally, it was test time, the clock was showing 12:45 as the Starshine team took their places behind their control panel. They had finally agreed to allow five remote cameras around the control panel for all the networks. It meant that the networks would all see similar pictures. One main camera would be angled slightly above and to the rear of the control panel. That camera would see the Starshine team from above and slightly behind, with a distant view of the test pad concrete walls, 75 yards away. The other four cameras were targeted head shots of the four team members and the control panel.

The count-down arrived. Denise pressed the big red 'Start' button. There was a snap sound and a faint humming sound for barely under six seconds. It was over. The light bar for the press and spectators blinked one time. All the gauges registered a 5.864 second sustain, and a meter showed that 160 megawatts (160,000 watts) of power had been transferred to the Los Alamos grid. They had produced 6000 times input. It was their best test of sustained fusion so far.

There was the usual 5 seconds of silence as everyone checked the dials and gauges, then a cheer went up at the panel. It was always an exciting thrill to see a test succeed but they were also playing for the cameras this time too. There were hugs all around and high-fives being traded. Needless to say, the test had been a 100% success.

General Wilson was there, just out of sight of the cameras most of the time. As they began hugging and throwing high-fives, he joined them on camera. It was a photo-op and they all knew it. The General needed to be in the picture. None of it would have happened without him. The girls both hugged him as he threw high-fives to each of the guys.

Next, everyone took a deep breath, off camera. They all swallowed hard, knowing what was next. Denise said, "Lets do this. There's no way out. We're a team. No cowards allowed." She laughed as she said it, but she really wasn't joking.

They all climbed into two separate cars and drove to the main gate where the press was waiting. Three different news choppers were flying overhead taking birds-eye shots of the scene, but the Press corps was setup at the gate. Every news association in the entire World had cameras setup in front of a large stage which was equipped with microphones a P.A. system, sprays of flowers and hoards of reporters. An MC for the base and the city of Los Alamos was already on the stage. He was keeping the air alive with ongoing banter about what had just occurred and what it actually meant to the World.

The team climbed out of the cars along with General Wilson in the lead. They followed him to the stage as cameras whirred and reporters began shouting questions. The MC was doing his best to calm the crowd but they weren't listening. Pandemonium had erupted as the Starshine Team gathered onstage. There were shouts, whistles and screams coming from the press corps and the hundreds of spectators who had gathered for the momentous occasion.

Finally, the MC shouted into the microphone and it rang loudly through the open air. "Here they are ladies and gentlemen, the heroes of the new millennium, The Starshine Team and Lieutenant General Frank G. Wilson. General Wilson is the man in charge of the Advanced Research Projects of the Army, Navy, Marines and Air Force at the Department of Defense. This project is his idea. You've just witnessed our tax dollars at work.

Let's hear it for General Frank Wilson!" Cheers and applause filled the air.

Frank stepped up to the microphone looking totally confident. "Well thank you all so much. Thank you, thank you, thank you everyone. You folks settle down now. Don't make me call in the National Guard!" Everyone laughed and the applause tapered off. "We are so proud of this bunch. I may have called them together, but they have done the work. They are five of the most remarkable people on the planet. They deserve the credit. First and Foremost I need to introduce this first guy.

This whole thing was actually his idea. He tricked me into giving a green light to all of this. As we sat talking one day a year and a half ago, he spelled out the dire need for a project similar to the Manhattan Project. For those who are too young to remember that project. It occurred in the early 1940s and gave us the Atomic bomb that won World War II and launched the Atomic age that followed. The Manhattan Project was the sole reason the U.S became the number-one, most prosperous country in the World as WW II ended. And it's the reason we are still #1 today. This project, 'Project Starshine' has just insured that we will remain #1 far into the future. But, this technology is not a technology that we can hoard to ourselves. We will be offering it to the World. It means that the Terror Wars over oil will end, immediately. The Global Warming issues are now 85% solved! We will have nearly free energy for many, many long years into the future! The man I want to introduce is the catalyst that caused it all. Incidentally, he volunteered for this project for the princely sum of one dollar per year.

This man is a visionary. He saw the promise of Atomic fusion and twisted my arm until I did something about it. When I finally did do something about it, it was apparent to me that he should be the man in charge. So, I offered him the job as top director of the Starshine Project. I need to tell each of you what happened next. We offered him $300,000 a year to accept the job. He refused every penny of the money! He volunteered to do it for the hefty sum of $1 dollar a year. Jim is a great friend and a great American. He deserves most of the credit here. He is the visionary who gave this project life.

How about a great round of applause for a man whose name remains top secret today. He is one of the top intelligence people at the DOD. We love him because he works cheap! Give him a big hand. He's the director of the Starshine Project, Jim Axe! Come on over here Jim!" Frank had stretched the name to sound like Axe on purpose because Jim X would have created too many questions among the Press corps.

Jim stepped up to the Microphone. "Oh stop it Frank. You're over doing it again. Don't kid yourself folks. One dollar a year is all I'm worth and they promised me two years in North Korea if I refused the job." He waited for a laugh that was lackluster at best. Maybe comedy wasn't Jim's best suit. He laughed at himself, a little embarrassed . . . and continued.

"I do nothing on this job and I'm worth every penny. The people who have done the work deserve the credit here. Let me introduce the leader of the Starshine project. She's not only beautiful but she has a Ph.D in Plasma physics. She's the leader of the Starshine team. This gal had begun preparing for Project Starshine, long before I mentioned it to Frank. This fusion idea has been her dream since she was in high school. Give her a warm welcome, Dr. Denise Feywisch."

A huge round of applause went up with wolf whistles and sophomoric catcalls. Denise was all smiles and bristling with excitement and inner beauty. "Thank you! Thank you so much!" Her violet eyes sparkled and she looked Hollywood beautiful as she beamed her gorgeous smile at the audience. "No one does a thing like this by themselves. It takes a team! Everyone of these people are the very best at what they do. Any one of them could take over my job in a second. Let's just bring them all out here." She called all their names one at a time as the rest of the team stepped onstage and up to the microphone with her. "These people are my heroes. We've been a team since our days together at MIT, and we've only just begun. This is just the first big test. Today's sustained fusion event was only six seconds in duration. The Wright brothers first flight at Kitty Hawk was longer than that . . . and look where aviation is today.

Fusion energy will be very similar. There is enough energy in the oceans to fulfill all our energy needs for billions of years. It is **ECSAFE**: **E**conomical, **C**lean, **S**afe and **A**bundant **F**ree **E**nergy! We're declaring the energy crises over as of today, and it will only get better with each passing week.

Other great minds will take this humble beginning we've made. They will take it to the stars and beyond, literally. Soon our space ships, our aircraft, ocean liners and all our homes will all be powered with **ECSAFE Fusion Power**.

Let me step back and let these guys,(the real heroes), have a word." First, this lovely gal here beside us is the REAL brains of this little group. She's an absolute math whiz. Audrey, is my best friend in the World. This project is almost entirely math, and very complex math at that. She solved the equations that made all this possible. Without her, we simply and literally wouldn't be standing here. We wouldn't have known where to begin without her. Dr. Audrey Reynolds! She' single, guys." The crowd erupted into more applauding, whoops, whistles and cat calls.

She hugged Gus around the neck and gave him a kiss on the cheek, then grabbed David around the neck with her other arm and kissed him the same way. "You've all seen MacGyver on TV. Well here they are in person; twice! We have two of them on this team! If it's broke, they can fix it! Dr. Gus Jenkins and Dr. David . Zimmerman! They're both single too, girls." A few screams and another round of applause followed.

Denise kept trying to step away from the microphone but the rest of the team kept pulling her back. Each one of the group took a short turn at the microphone with a lot of smiles, bragging on each other and tossing a few corny jokes that fell pretty flat.

Finally Denise took over again, "We're about to wrap this up, but we know that you people in the press have a lot of questions, so we'll try to answer a few before we get out of here. Ask away!"

The team took turns answering the easy questions. When someone asked a sensitive question, Denise would step up and say "Sorry! That's top secret. We can't talk about it."

After about 10 minutes of questions, Denise called General Wilson and Jim back to the microphone with a smattering of lavish praises and happy smiles. The Starshine team left the stage waving to the crowd and cameras to a thunderous round of cheers and catcalls.

Within another two or three minutes, Jim and the General turned it all back over to the MC and they too left the stage.

The Starshine team piled into their two cars and headed for the house on Arizona avenue.

As they turned a few corners they realized that a line of traffic was following them. They couldn't go home! They would never get rid of the crowds if they went to the house on Arizona avenue.

Denise came up with the only thing they could do on such short notice. She called Audrey in the other car on her cellphone and yelled, "Audrey, go to the lab. We need to get behind the gate where the crowds can't go. We'll never get rid of them if we don't."

The entourage made a few more turns and headed back the way they had just come as the news choppers buzzed overhead. The two car loads of celebrities breezed on back through the security gate and into the restricted area. Running from fans was a new experience for all of them. It was over for now. They would simply chill-out at the lab until the crowds were gone.

When they arrived at the lab; the director of the lab was ready and holding open the door, with a smile.

A happy and smiling General Wilson handed him a fifty dollar bill and begged him to run through the gate and bring back two cases of beer, iced down, for the team. The director left immediately in his own car, happy to do the small errand that would have been impossible for anyone else in the group.

The General grinned, "We might as well have a little celebration right here. "You are all big-time celebrities now. You can't even go home without being mobbed on the way. How does it feel? It'll calm down after a little while. You folks have put this old town back on the map."

Two ice chests full of beer arrived a short time later. Everyone clicked bottle necks and toasted the future.

Denise checked her email as they all sat around relaxing with a beer. A mysterious encrypted e-mail from her brother Eddy at the DOD appeared. It was coded in the cipher they had used as kids; growing up with computers. It took her 30-minutes to decipher it by hand. It read:

Sis,

Rachel has been snooping the computers at the NSA again. Her job puts her in a unique position that gives her direct access to private e-mail of certain specific upper-echelon directors at the NSA. She is seeing a plot that needs to be investigated by General Wilson at the DOD. We can't inform him without divulging Rachel as the source. They would surely hang her for snooping.

It appears to be serious. There are several of the higher echelon people at the NSA who are invested heavily in oil related stocks. Your project is causing all their stock holdings to fall towards precarious lows. You are a threat to their bottom-line. They are very aware of what you people are doing and they are conspiring to stop you by destroying your reputation. They do not want to see fusion replace fossil fuels. Some of these people are apparently big players within the fossil fuel industries.

Rachel is seeing inordinate amounts of e-mail traffic back and forth from the NSA. The e-mails are moving to and from a consortium of oil conglomerates in Houston called GULCON. They are plotting together to bring Project Starshine to a tragic demise.

Their objective appears to be a multi-pronged attack against you. They are preparing an expose' of the black budget funding the DOD has received for your project. They are following the money. They know that the black budget project was granted $20 million dollars and that you have nearly accomplished the job for less than $3 million . They are planning to go to the press with allegations of inappropriate black-budget skimming by you people, and the DOD. They're following the money and looking for ways to destroy your credibility. They are also wanting to bring down General Frank Wilson and replace him with a different, more oil friendly General who will cancel your project as he takes the position. Apparently they've discovered that he has a wondering eye. They mentioned photos of an affair he had with a young lady ,a clerical worker at the Pentagon.

According to the encrypted e-mails we are seeing, they are planning to go after you several ways. After they have destroyed your credibility they are also planning another hack attempt on your computers. The really frightening thing we saw that followed, was a simple one line remark in one e-mail, that stated: "**If we can't stop them any other way, we will be forced to eliminate the brain power**."

General Wilson needs to be aware of the plot and all of you need to account for every penny, and make it public. Change all your passwords and the URL paths to your files. Store backups of the diagrams and equations off the computer and start watching your back everywhere you go.

Find a way to get this news to Wilson without implicating Rachel or me. We are breaking some very serious departmental rules by telling you this. Our actions could be labeled as espionage because we are deciphering private e-mails of some *very important people. Those people have the power to send us to prison forever, . . . if they don't have us executed for treason.*

You need to put your genius IQ's together and find a way to alert General Wilson before your enemies have a chance to drag your project through the mud or harm you physically. ~ Ed

As the team sipped beer and relaxed, Denise went to work on the computer. First, she downloaded all their file updates to flashdisks. Then as a stroke of genius, she filled the current site with folders and files that contained dummy files, the equations and diagrams that had failed. They appeared to be functional working diagrams and equations, but they were all flawed. If they were pilfered, the thieves would think they had the working files, but in effect they would have only the failures. It would give them a false sense of 'mission accomplished'. She left the password and URL as they were for those files, knowing they were vulnerable.

Then, she setup an entirely new digital storage area that appeared innocuous. It looked like an unimportant digital storage area with pictures of someone's grandchildren and pets. It was protected with a new encrypted password. The new storage area contained all the real and functional folders of diagrams and equations. With that done, she breathed a sigh of relief. It would take a a team of genius hackers to even find the folders, much less break the new password.

Finally, after several beers, the team decided it would be safe to go on back to the house on Arizona. The crowds had given-up and dispersed. It seemed calm enough to go on home.

Once Denise was in the Mustang alone with Jim, she told him about the ominous e-mail from her brother. She asked him to help her find a way to alert the General without revealing Rachel and Eddy. The two of them sat in front of the house a few minutes, trying to devise a plan that would reveal the plot without implicating her brother and his girlfriend.

Finally they went on inside and the General called Jim aside. "Jim! Some bad omen's appear to be in the wind. I have sources and friends within the NSA and I just got a call from them. They seemed to have uncovered a plot that appears to be threatening. We need to nip it in the bud."

Jim was stymied for a second. It appeared that the General already knew what he and Denise were worried about. He decided to hold his tongue. "What's going on, Frank?" If the General already knew of the plot, it would make it a lot easier and safer for Eddy and Rachel.

Frank Wilson described the exact same plot that Eddy had warned about. Apparently the General had his own sources inside the NSA as well. That made everything easy.

Jim simply listened quietly. That's all he had to do. Together, they devised a plan to put the financial books straight. They would keep an extra $1 million of the black budget money for additional and excess funding, but they would return $16 million directly; with President Clayton and several other top officials as witnesses.

The General had put his own team of Departmental Intelligence on to the GULCON consortium. They were locating and making deals with fringe people on the inside of the organization to gather info on specific key individuals who had come to his attention via his secret sources. They were gathering evidence that would put the key NSA people in jail.

The General then called the team together and alerted all the members of the Starshine team that their lives might possibly be in danger. A short phone call later had a special squad of Army specialists dispatched to guard the Starshine team. They would arrive the next morning. He explained to the team that they wouldn't know the guards were watching. They would never see them and they wouldn't get in their way, but the guards would be keeping an eye on them 24 hours a day, for awhile. The General warned that they were being followed and watched for their own protection. There was ample due cause to worry for the safety of the team."

Jim breathed a sigh of relief but It was unsettling to know they would be followed and watched at all times. However; it was good to know the DOD would be looking out for them. It would make them feel a little safer.

General Frank Wilson flew back to Washington DC the next morning.

Chapter 23: Enrique: Caught!

Agents of the FBI were hot on Enrique's trail. They had learned a lot more about him. He was not all he had claimed to be. The story about being a descendant of the Mexican Revolutionary, Zapata, was true. And, it was true that he was an heir to millions of dollars on his mother's side of the family, but he was not a legitimate citizen of the United States.

He had grown-up, living with distant relatives, using their last name of Smith since the age of 7. The family money had created a fake identity for him when he was very young and he had grown-up in New Mexico as a normal American kid. That explained the impeccable American accent, but he was actually an illegal Cuban alien.

On being accused of a federal crime, he knew it would all come out and he knew that the FBI would eventually find him and most likely deport him. He had but one sane choice. He would make a run for Cuba.

Agents caught him boarding a private plane in Miami. After several intense hours of interrogation, they automatically assumed he was guilty of espionage against the United States. This meant he was then classified as a terrorist. Since he was actually a foreign national, it meant just one thing: he was not guaranteed a fair trial in the U.S. courts. He had no legal rights in the U.S.

He was labeled "Terrorist!". It was stamped across the top of his FBI file. He had been labeled with the crime of espionage and had been instrumental at trying to subvert a top secret military project. That crime qualified as a terrorist-act against the United States of America.

Since Enrique was a Cuban national, it meant one thing to the Generals at the DOD: GTMO! They would send him to Camp Delta for an undetermined incarceration at the Guantanamo Bay detention camp facility in Guantanamo Bay, Cuba.

It was a fate worse than prison! The people at GTMO made their own rules! That detention camp was controlled entirely by the military. GTMO had been criticized in the past by many human rights groups, for human rights violations of every sort, including water-boarding and several other debatable forms of torture. It is said that at least three detainees had died there in the past while being tortured. GTMO was not bound by US laws and the detainees weren't guaranteed the same rights as the people who were incarcerated in U.S. prisons. GTMO was a nightmare existence and there was no term-limit placed on the length of the detainment. Enrique could and very well might remain there the remainder of his life. It was the worst fate imaginable. He would simply disappear and never be seen again. Case closed!

A laughing and exuberant General Frank Wilson called Jim and told him the news. He saw it as a fitting penalty for a guy who had so blatantly betrayed the team. Jim expressed the opinion that it seemed too harsh, but the General disagreed. There wasn't a lot Jim could do.

Jim had to relay the information to the team. He dreaded telling them. He knew it would hurt Audrey even more. She was angry and she had been betrayed, but she loved the guy. She would never have wished for that sort of misery to become his fate. She burst into tears and ran from the room as Jim gathered the team to tell them. They all agreed that the punishment seemed to be unjustly harsh.

It was what it was! There wasn't anything any of the Starshine team could do. Jim had tried to convince the General to take a different stance but the truth was, the military didn't want to go through a lengthy legal trial. It would divulge too much about the top secret project. He did say that once the project was concluded; they might cut Enrique some slack and turn him loose to Cuba.

Gus was almost as upset as Audrey, concerning Enrique's treatment . He had already suspected that Enrique might be innocent, though he hadn't voiced that suspicion to anyone on the team. No one else seemed to be suspicious of Brittney but him. They all assumed she was a little too slow-witted to do anything of that sort.

It had occurred to him that though Brittney McGill didn't seem to be bright enough to pull off an espionage scheme of that nature, it wasn't totally out of the question. She aspired to become an actress and she was beautiful enough to turn heads in Hollywood.

What if she had been playing him and playing dumb all along. What if she was acting her way through life? It was a possibility.

Another thing was bothering him. Brittney had recently been too busy to talk to him. He hadn't been overly concerned about it and he had only called twice. Gus kept a little black book with the phone numbers of numerous other potential young beauties. Brittney wasn't even at the top of the list. The affair with Brittney wasn't a love affair. It was a lust affair.

But, she was never available when he called. That seemed additionally curious. She lived with a group of friends in a small apartment across town. When Gus called, they wouldn't say where she was. She was just never home. Gus decided to do some detective work.

Around 5:00 PM one afternoon on his way home from the lab he decided to drop by her apartment. As he parked in the driveway he noticed that a white, late model Chevy pulled to the curb several blocks away. He hoped it was the Army surveillance team. It was obvious that he was being watched and followed by someone.

He rapped on the door and Brittney's friends invited him in. They were all there, but Brittney wasn't. Gus feigned that he really cared for her and he was really missing her. He begged them to reveal where she was and slowly won them over. They were becoming sympathetic. They were all a little starstruck with the fact the he was now a celebrity.

Finally they told him that Brittney had recently come into a lot of money and she had booked a Caribbean Cruise. She was gone. She had taken all her things, moved out, and did not plan to come back. She had told her friends she planned to live in West Palm Beach, Florida after returning from the cruise.

Brittney had made her friends promise to not reveal where she had gone, but Gus was doing some acting of his own. He convinced them that he and Brittney had simply encountered a little lovers-spat. It had been an insignificant, silly fight. He really loved her and missed her. He persuaded them that if he could just locate her, he could win her back. It was all very convincing and he sounded so romantic. They fell for it.

He won all of them over! They all saw him as a super hero anyway because they now knew he was a famous physicist with the Starshine Project. They all considered Brittney stupid for running away from someone like Gus. He was actually out of her league and they all knew she wasn't playing with a full deck. So, in her best interest, they told him the name of the Cruise-line and they even gave him the address of the new condo Brittney had recently purchased in West Palm Beach, Florida.

Gus decided, rather than subject her to the scrutiny of the DOD and the FBI, because of his uncertain and unfounded suspicions, he would simply surprise her in West Palm and discover the truth for himself, first-hand.

He went home and called a meeting with the entire team, then told them exactly what he was thinking, explaining that he wanted to take a few days off to investigate the matter personally. He was almost certain that Brittney had betrayed them, and not Enrique. He felt a moral duty to free Enrique, if he could discover for certain that Brittney was the guilty party.

Jim was immediately on the same page with Gus. He had seen a certain look in Brittney's eyes that day at the pool. She had looked like a deer in the headlights. It looked like fear, surprise and bewilderment. She seemed to be shocked at the fact that Denise had found the computer breach so quickly. She had only displayed that look for a second, then she recovered and continued playing dumb. It was very possible that Brittney was the spy and not Enrique.

Jim laughed. "My, my! How convenient that she should choose Palm Beach, Florida. I just happen to own a vacation home there and it comes complete with a limo, a chauffeur and the lavish lifestyle of the rich and famous. I'm going to propose a working vacation for all of us . . . and I don't want to hear any arguments! We all deserve this!

I want everyone on the team to think about this wild idea I'm having . . ." He explained his idea in such a way that everyone went along with it. It was a terrific idea! Denise hated to take any time off from work but she agreed that the break was warranted and the idea was worthy of the trip.

A quick check of the Cruise Line proved that the Cruise ship Brittney was on, would dock at Fort Lauderdale, Florida the following day.

Jim had the Learjet flown in and he would pay for the entire working vacation to Palm Beach. He called and put the chauffeur and staff at the Palm Beach mansion on alert. He told them the entire Starshine team would be arriving the next day.

Chapter 24: Confrontation With Brittney McGill.

The Learjet arrived in Los Alamos from DC the next morning. Since there was no rush, Jim convinced Gus and the anxious team to wait an extra day before leaving. He wanted to give the pilot a little time to rest before making the additional long flight back across the country to West Palm Beach.

After the pilot had rested for a day, the team flew out of Los Alamos the following morning, arriving at West Palm International Airport around 4:00 pm in the afternoon. The limo was waiting on the tarmac. The smartly dressed chauffeur literally ran from the limo to escort everyone from the plane to the limo, and he quickly loaded their light luggage into the trunk. He then climbed behind the steering wheel and hit the intercom. "Where to, sir?"

Jim hit the intercom button and replied, "To the Palm Beach estate Charles. Thanks."

Denise was thoroughly impressed at this point. They all knew Jim had money but they had no idea that he was a part of the Palm Beach upper-crust elite. He just didn't carry himself as if he had that much money.

As the limo rolled through the big gates and into the estate, there were a lot of oooohs and aaaahs. The estate was enormous and fabulous. It sat majestically on a raised hill, overlooking the beautiful Atlantic ocean. A warm, gentle breeze swayed the palm trees and shrubs in the yard, adding an inviting tropical feel to the setting.

As they arrived, Jim showed everyone around the house and he showed them the bedrooms where they would be sleeping. He had asked the chauffeur to hang around because there would be another pickup and delivery across town.

Gus, Jim and the chauffeur had worked out a spectacular plan for Brittney. It was a complete ruse.

The Chauffeur and Gus left a while later. Gus had the address for Brittney's new condo. It would be a surprise visit.

The Chauffeur waited in the parking lot as Gus took the elevator up to Brittney's condo address. He rapped anxiously on the door.

Brittney was in total shock as she opened the door and saw Gus standing there. At first she started to close the door in his face, but it was too late. Gus had his toe in the door and he pushed his way in, smiling with all the charm that was typical Gus. Brittney was actually glad to see him. He had won her heart long ago. She hadn't left him because she was mad or because she wanted to get away from him. She had left because she had to leave.

Finally, after a tense moment, she threw her arms around his neck and planted a wet open-mouthed kiss on his lips. Before they knew it, things were about to go ballistic.

Gus pulled away. "Brit! I have something special to show you. There's a lot you don't know about me. I want you to come with me over to my place. We have a lot to discuss. Will you come with me? I have a really nice place here. I want to show it to you."

"Sure. Okay. Why not?" Brittney was a little curious but deep down, she felt guilty and she was secretly hoping that she might mend things with Gus. Somewhere in the depths of her mind she hoped the two of them might live happily ever after.

Gus escorted her down the elevator to the waiting limo. Charles, the Chauffeur held the door for them. Though it was prearranged, Gus told Charles to take them to the estate. He added an little extra tone of superiority. It was all a fake effort to impress Brittney.

Brittney's eyes were like gorgeous full moons. She had no idea what was happening.

A few minutes later they pulled into the circle drive of Jim's estate. Gus hit the intercom. "No need to get out, Charles. We've got it from here. That will be it for today. Thanks."

He opened the car door and led Brittney up the steps to the tall and ornate, hand carved double doors.

"D-do y-you live here?" Brittney asked timidly. This is some of the most expensive real estate in the World.

Gus just smiled. "This ole shack? Sort of, for the time being anyway. Come on in!" He opened the door to the dark mansion.

Immediately a light went on and everyone there screamed, "Surprise!!!"

Brittney nearly went into shock. She screamed at the top of her lungs, then in a brief second as they all watched her eyes and her face, she showed nearly every emotion known to mankind. Her face said it all. First there was shock, then happiness, replaced by guilt, then remorse and then fear. The final look stayed on her face, as she caught on. It became a look of abject fear. She looked at Gus, who was glaring angrily now. They were all glaring angrily. She instantly knew they were on to her, and there was no apology or excuse in the World that would explain what she had done. The poor little dear really was in the headlights this time, and an 18-wheeler was bearing down on her at full speed.

Gus spoke first. "We want to hear the entire story Brit. Why, how and who? Spit it out!"

Brittney turned to face the door as if she might make a run for it but David and Denise stepped in front of it with their arms folded across their chests.

Audrey was livid. She wanted to tear into Brittney physically and scratch her eyes out. She had been through a horrible Hell because of her. Enrique was still going through a more unimaginable Hell because of her.

Jim stepped in before things got out of hand. "Brittney, we want to hear your side of the story. We could have had you picked up by the FBI, but this may be a little kinder. Start at the beginning and tell us everything. If it's good enough, we may try to get the DOD to cut you some slack. You are in very serious trouble little lady. Tell us why you did it . . ."

Brittney swallowed hard and then she began mumbling through her very real sobs. "Money! It was only for the money! It was so much money! I've never seen so much money as that. I knew I would never, ever have an opportunity to make that kind of money again in my life. I thought I could pull it off.

Honestly, a few silly computer files didn't seem like a very big deal at the time. It seemed trivial compared to the payoff. I didn't know it would ruin the project. I figured you guys would recover in an hour and I would be $5 million dollars richer. They paid me $5 million dollars just for the user name, password and the uniform resource locator path to your files.

So . . . one night after Gus went to sleep, I went through his wallet and there it was, all in one place on one little piece of paper. It was so easy.

Some very strange and mean looking men I had never seen before, approached me at work. They didn't say who they were or who they were working for. They gave me $10,000 up front after I agreed to try.

I was already dating Gus at that time and they knew it. That's why they came to me. They figured it was the easiest way to get the passwords and the URL. I guess they were right. I was able to get it on the very first night. When I looked for it the first time, I found it in . . . like ten seconds. She grinned sheepishly at Gus and said, "You really should be more careful, Gus!"

I simply asked myself where it might be, and where a guy might keep something like that. I immediately went for the wallet while Gus was sleeping. There it was, in among several $100 bills. I didn't touch a penny of the money, but that $5 million was calling my name! I feel just terrible. I love all you guys. You are all great people. I fell madly in love with this hunk here (Gus) but I knew I was only a toy to him. He just isn't ready to be serious with anyone.

So . . . the money, the money . . . the money! That's why and how. I still don't know who I did it for. They remained anonymous. They didn't want me to know who they were and I didn't ask. They had given me a number to call if and when I had secured the password, and all that. The number was a throw-away phone, I'm sure.

I called and told them I had the goods. They brought two inconspicuous looking large heavy duty garbage bags full of $100 dollar bills. Cash and unmarked! I've never seen so much money in all my life! After they had shown me the lovely contents inside the bags, I gave them the passwords as we simultaneously made the switch. They disappeared in one direction and I went the other. I have no idea who they were or where they went." Brittney was crying now, actually sobbing . . . and she wasn't acting. To be honest, I did not trust them so I had them meet me in the parking lot at CVS in broad daylight. I knew there would be a lot of people around. I was still afraid, but that kind of money will mke almost anyone do crazy things. They paid me in full!

Everyone on the team felt a little sorry for her. As they thought about it, it wasn't hard to imagine anyone doing a little job like that for $5 million dollars. That's a lot of money. Most normal people would never see that much money at one time, in their entire lives.

But, all this meant that Enrique was languishing away in the detention camp down at GTMO and he was completely innocent. His private Hell wouldn't be over until they got him out. They would have to turn Brittney over to General Frank Wilson and she would have to confess to him and suffer the consequences.

Even then, Enrique's future was still bleak because he was an illegal alien. They might turn him loose but they most likely wouldn't allow him to re-enter the United States.

It didn't matter. They simply couldn't go easy on Brittney. General Wilson would have to decide this one.

Jim was on the cell in seconds. He hit the speed dial to connect with General Frank Wilson. The General picked-up on the 4th ring. "Jimmy, old buddy! What's happening?"

Jim was somber but he faked a brighter greeting than he felt inside. "More earth shaking news Frank. Sit in a chair if you aren't already sitting. This may be hard to take."

Jim chuckled, but only to add a small amount of levity to the conversation. "Enrique is completely innocent. He didn't do it. We have proof sitting here in front of us. He's a patsy. He was beautifully and accidentally framed as a matter of being the wrong guy, in the wrong place at the wrong time. He sort of framed himself, much to the good fortune of the real traitor. The real traitor is Brittney McGill, Gus's girlfriend. I'm sure you remember the string bikini at the pool. We are with her right now. She just confessed the entire ordeal. We're even a little sympathetic to her situation. What she did was stupid, but 'traitor' and treason is a little bit of a stretch, Frank. Greedy is a more apt term. The boys in Houston paid her $5 million to pull off the tiny caper. That's enough money to tempt Jesus himself, as we're seeing it. We're hoping for two things. First and foremost Enrique needs to be freed immediately. He's 100% innocent. Secondly, we hope you'll help us find a way to go easy on Brittney. She did it for the money, plain and simple. A lot of money! Any of us might have done the same thing if we were in her shoes. $5 million is a lot of money for just a password and a few computer files. Be thinking about that.

Meanwhile .. . what can we do about Enrique. We all owe him a huge apology. He has been wronged in the worst possible way." Jim was doing his best to rectify the situation for everyone concerned.

"Whoa! You really know how to make a General earn his keep, Jimmy my boy. This is teeth rattling stuff! Are you sure you are right?" There was still some doubt in the General's voice.

"Yes Frank. We are all 1000% certain. There is no doubt. Brittney is sitting right here. We became suspicious and we flew to where she ran. We are all in Palm Beach Florida right now. She is sitting next to me on the couch. Would you like to speak to her?" Jim looked at Brittney. She was still sobbing uncontrollably. It was hard to watch.

Frank went silent for a few seconds. "Is that her I hear balling and blubbering her eyes out in the background?"

"Yes it is." Jim responded.

"Well, she's in no shape to talk to me. I believe you anyway. I'm on it. I'll be on the horn to GTMO in the next few minutes. Would you want to fly down and get Enrique tonight? I figure you have your little jet parked there at the back door." He laughed because he knew Jim well enough to poke a little fun at his wealth.

Jim considered it for almost 2 seconds. "I hadn't thought of that Frank. Yes! We'll fly down tonight and get him, if you can get us the clearances. We'll fly in, pick him up and bring him back with us. That would be just great! There's a pretty little red head standing here with me who would love you forever, if you can arrange that for us. We owe the kid big-time, Frank. He's been through Hell and you know it. I think Audrey may faint any second. She's about to come apart with excitement. This is all a bit much for her. She loves that guy. I'm betting we will see them get married before the end of the year."

Frank sat silent for a second, then spoke. "You're right. I'd say we really owe Enrique a huge apology. I'll tell you what. He's a Cuban national. I can fix that. We'll make him a citizen if he can forgive us. Tell him that, as you pick him up. He'll be treated like royalty after I make a quick call down there. You can pick him up as soon as you can get there."

Jim wanted to use every ounce of pull he had with Frank. He hated to see Brittney go to prison for doing what 999 out of 1,000 people might have done in the same circumstances. "So, that will solve the Enrique problem. That's great!

What do we do about Brittney?" Jim looked at Brittney, smiled and held up his crossed fingers. He was hoping to get her off with a hand slap.

Frank took a few seconds. "She actually screwed $5 million out of those creeps in Houston, just for a password and a few computer files, that they didn't get to keep? Good for her!" He laughed. "Jeesh! I might have gone for that myself. The hard part would be persuading Gus to sleep with me!" He laughed heartily at his dumb joke.

"Okay. I'll tell you what. I can tell you are all sympathetic to her situation. I am too, a little, and we have the files back. The boys in Houston have lost $5 million and they got nothing in return, the way I see it. I'm real happy to see her run away with their money. It seems to be a fitting touch of irony. I'm okay to let her go if you guys are. I don't want to prosecute that pretty little thing. I can remember that she is a gorgeous eyeful. We'll let her slide. How does that sound?"

Jim brightened up and tossed a smiling wink and an upturned thumb at Brittney who was now sitting with both eyes wide, in total terror. Up until this very moment she had thought that her life was in complete ruins.

Jim laughed. "Well, Audrey may not love the idea but I think the rest of us will sleep easier with your solution. I'll tell them; we'll take a vote and see what happens. Audrey may want to go a kickboxing round or two with her one-on-one, but the rest of the team will be okay with it, I think.

Get us clearance to GTMO. We'll go get Enrique tonight. I don't want him to languish in that place another minute. Audrey will be thrilled with that. I'll go now. I need to alert my pilot and get the Learjet ready to fly out in an hour or so. Thanks Frank. You've just proved to me why they put those stars on your shoulders. Well done my friend. I'll talk to you soon. Bye." They both disconnected at the same time.

Everyone had heard Jim's half of the conversation and they could tell from his spirits alone that it was all good. He took a few minutes to explain what the General had said.

Audrey had already heard half the conversation but it was priceless to watch her eyes as he explained that they would be going to pick-up Enrique within the hour. She glared a little as he explained that Brittney was to go free and keep the $5 million. Brittney was no longer crying. She was in absolute shock, but she was afraid to show too much of her obvious happiness.

Audrey was cool concerning Brittney. Eventually she admitted that it was sweet irony that Houston had lost all the way around. And . . . she was exuberant that they would be leaving in a little while to pick up Enrique. She was happier than she had been in weeks.

Everyone talked it over. They all agreed that Brittney should not be on the plane or anywhere near Enrique. It just wouldn't be safe after what she had caused him to go through.

Gus called Jim away from everyone else and explained that he wasn't angry with Brittney and would like to bunk with her one more night before they headed back to Los Alamos. Jim just smiled. He understood perfectly. Gus called for a cab. He and Brittney left for her new condo a short while later.

Jim called the Chauffeur. He was embarrassed to call him at such a late hour and on such short notice, but the guy was on a retainer and being paid very well to be on call. Jim explained the situation and within thirty minutes the limo pulled up out front.

After alerting the pilot, David, Audrey, Denise and Jim piled into the limo 45 minutes later, drove to the airport, climbed aboard the Learjet and in under two hours the Learjet was touching down on the runway at Guantanamo Bay, Cuba. They taxied slowly up to an empty terminal where a lone figure of a man stood under a dim light, holding an overnight bag.

Audrey was bursting at the seams! "Open that door and get those steps down or I'm gonna put a hole in the side of this airplane!" She yelled. The startled pilot quickly did as she had asked, with a grin. She flew down the steps and into the arms of the only love she had ever known in her life. Enrique dropped his overnight bag to the tarmac and grabbed her, then swung her around. The two lovers kissed passionately for what seemed like hours with everyone watching. Jim, Denise and David had gone to the front of the plane where they could see, and they were all watching moon-eyed and smiling, through the windshield.

It seemed to take forever, but the team was very patient. They all stayed on the plane because Audrey and Enrique (aka; Manuel) needed their special moment alone together. It was their one very special moment in time. We all have a moment like that somewhere in our lives. It needed to be just the two of them . . . alone in the moonlight, on the tarmac. (And . . . it was storybook perfect.)

Soon the two lovers came up the steps of the Learjet. The pilot closed the door, taxied into position, waited for clearance from the tower and seconds later they were back in the air.

Audrey and Enrique cuddled and necked all the way back to West Palm Beach.

Two days later the entire group went back to Los Alamos, without Brittney. Gus was smiling. Brittney was fun, but he doubted if they would ever see each other again. There just wasn't enough there for Gus, the playboy.

Los Alamos wasn't finished with the Starshine Project and the Starshine Project wasn't finished with Los Alamos. Things were about to get really good.

Chapter 25: Building the ultimate Dream!

After the Starshine Team returned to Los Alamos, they resumed working on the reactor. There was still a lot to do. It would take at least another year to perfect everything and get setup to go public with a commercial reactor model. The days quieltly melded one into the other for awhile.

One afternoon after work, Denise decided to stop by the grocery store on her way home from the lab. Italian food was her favorite. She had decided to go home and cook a big meal for everyone. Her taste buds had been craving Linguine all day long. It took 20 minutes to buy everything she needed for the meal.

As she headed back to the Mustang with the groceries, she noticed that a suspicious van had pulled into the parking space beside the Mustang. She placed the grocery bags into the front seat on the rider's side and was about to climb behind the wheel when she realized someone; a very strong man, was behind her.

Instantly a rag that smelled sweetly rancid, was cupped over her nose and mouth. Her World turned upside down almost immediately. She passed out. Everything went dark. The last thought that went through her mind was, "Oh my God. This is it. They've done it. I may never wake up from thi . . . ". Denise was out cold.

Hours passed . . . she was dreaming, seeing strange visions, odd images of people she didn't recognize. She was hearing strange and different sounds. She was coming around. Finally she opened her eyes. She was in a hospital room.

Jim was at her bedside. "You're in the hospital Denise. Everything is going to be okay. You were nearly kidnapped.

Luckily the Army surveillance team was nearby and watching. They caught the people who were trying to kidnap you. Those people are being interrogated right now.

It's almost 9:00 PM. You've been out for over three hours. Did you sleep well?" Jim was holding her hand and smiling. He pressed a button near the bed to call for a nurse or a doctor.

"Whoa! What happened? One second everything was fine and the next second everything went black." Denise raised on one elbow. "I'm a little dizzy. What the heck happened?" She was puzzled. Only small fragments of memory were helping her. She remembered being in the store . . . then walking out . . . and everything went black. "What did they do to me?" She asked.

Jim smiled confidently. "Nothing. They didn't have time to do anything. You were saved by the Army surveillance team. Those guys the General had guarding us, saved you. They caught the people who accosted you. They caught them red-handed and the thugs were caught and apprehended right in the middle of their intended caper. We've got them dead to rights. I think it will lead to convictions of the big Houston people behind all our nightmares. Your potential kidnappers were obviously hired by the GULCON consortium people. As they're interrogated, they may lead us to the BIG people who want us out of the way.

A doctor walked in and said hello to both of them. He grabbed a chart at the foot of the bed and read over it. "We think they used some strange new formula to knock you out. How do you feel?" He asked.

Denise shrugged her shoulders, "I was a little dizzy but my head is clear now."

The doctor smiled sympathetically as he took her pulse and shined a light in her eyes. "I think you will be fine. Your vitals are all perfect. You can probably get dressed and go on home if you feel okay." He looked a Jim. "Will you be with her the rest of the night?"

Jim responded, "Yes."

"Keep an eye on her. If she starts acting strange, passes out or anything, bring her back in. She seems okay to me. I wish I knew what they used to knock her out. It's new to us. We couldn't find a trace of anything abnormal in her system. It seems that the bad guys are always a step or two ahead of us.

If the interrogators can find out what they used, please come back and tell me. Your clothes are here in this bag, ma'am. You are free to go. Get in touch immediately if you experience any strange symptoms. I think you'll be just fine." He turned and left the room.

A short while later Jim drove Denise back to the house on Arizona avenue. His cell was ringing as they stepped through the door.

It was the Army surveillance team leader. The three men who had tried to kidnap Denise were singing like canaries. The DOD had offered them deals to save their own skin. The General at the DOD wanted fingers pointed to names inside GULCON. He wanted to nail the leadership. We've got some pretty big people on the run. It will all be handed quietly. A few of those people may soon be taking an extended vacation to GTMO. I don't think they will bother you again. Just the same, we will be watching. We aren't going anywhere, just yet.

Days and weeks passed. No other threats occurred. Progress with the reactor was moving along nicely. Within two months the team was getting good results on an hour long sustained reaction. They had pushed the electrical output to industrial levels and they had done it all within one plasma containment vessel.

Just for kicks, Denise began playing with her ideas for sequencing the plasma containment. The concept was similar to bouncing a hot potato from one hand to the other.

She drew up a design and called the team in to tear it to smithereens. To her surprise, there was more agreement than she had expected. Gus and David thought they could build a trial version fairly quickly. The trick seemed to be in the design that allowed the plasma to remain stationary and the containment vessels rotated and snapped quickly into place in a microsecond.

Gus came up with a device he called a shuttle. It maintained the delicate balance that was essential to perpetuate the fusion reaction, but it shuttled the containment vessels successfully from one containment vessel to the next, quicker than an eye blink. It was integral to Denise's sequencing design. Gus became the co-author and co-inventor of that patent. Everyone thought they should give the concept a try. It might lead to much longer containment and exponential output .

Once the initial fusion reaction had occurred there was no need for an additional input of energy. The system would become for all intentional purposes, self-sustaining. In a sense it would become in effect a self perpetuating something or another. No one wanted to call it perpetual motion; but essentially, that's what it would become, as long as the small amounts of fuel were consistently available.

Shifting the reaction back and forth to different containment vessels, would increase the output and efficiency of the entire project exponentially.

Also; if they dared to dream really big and if the idea worked above and beyond their expectations, it could lead to unlimited energy production. The door would be opened to the possibility that fusion reactions might be sustained for unlimited periods and energy on demand. It would also help them find ways to control the amounts of energy being produced, so that the electrical generation system could be increased or decreased depending on the loads and the demands. If they could do all that, the World would beat a path to their door. It was a very far-fetched dream but it was just the sort of the dreams that had brought them this far and it was the sort of dreaming that has consistently made America great.

The experiments continued as weeks turned into months. They tried the sequenced containment concept and it worked. By March of the following year they were sustaining the fusion reactions for unlimited periods. They developed the reactor to sequence between four different plasma containment vessels. Amazingly enough, the team discovered ways to control the output depending on demand and they had gained the ability to sustain fusion 24/7/365 if they wished. Model SL-a1 was ready to go into production. They lovingly called their creation 'The Slave'. It would work endlessly and tirelessly for only the cost of initial ignition and about two pounds of Tritium and Deuterium per month, which was minimal. Sustained Fusion had become a reality.

The Starshine research team had applied for patents on the reactor and the patents had been approved. They submitted papers to the proper institutions that explained in detail what they had done and how they had done it. All of the Starshine Project details were now appearing between the covers of the World's most esteemed Science journals and magazines. They had made the covers of most of those magazines. Some 4,000 web sites now covered the topic of sustained fusion in minute detail and the Starshine Project was the primary focus of virtually every one of those web sites. Sustained fusion had become the new phrase on everyone's lips.

The television talk-show circuits were all calling and hounding them daily, to appear. No one on the team actually wanted to do it. They just weren't a publicity-seeking bunch. All four tended to lean towards introversion. They really didn't see themselves as star material on prime-time television. They were basically thinkers and quiet, intellectual types. The fact that they had balked and said 'no' so many times was causing the offers go up and up. They were actually the biggest news in the World at the moment. Everyone was clamoring to know what fusion would mean to the future of the world and everyone was especially anxious to see the people who were responsible for changing the World so drastically.

Finally, one Monday morning Ellen DeGeneres called, personally. Ellen's secretary had called Jim's cell phone and asked to speak directly with Denise. She explained to Jim that Ellen DeGeneres wanted to speak directly with Denise.

Jim handed his phone over to Denise as she worked on the reactor. He had both eyebrows raised and he fairly well ordered Denise to respond. It's Ellen for God's sake! Take the damn phone!

Ellen came on the line, using all her charm. Before it was over, she had offered them all $50,000 each for a one-hour appearance. She was willing to shell out $250,000 for all five members of the team, if they would come on her show for an exclusive in-depth interview. She wouldn't accept 'no' as an answer. She ended up asking how much it would take, to get them on the show. It was as if she was offering a signed blank check. They could name their price.

Denise was shaken, just knowing that she was chatting on the phone with one of her own personal heroes and being offered a blank check!? That was too much to resist.

"I'll call a meeting and see what the others say. If they will all agree to do it. I will. Give me a day to work on this." She was literally trembling as she hung up the phone. It wasn't the money. It was Ellen: one of her most favored people in the World!

Soon the team was in a huddle tossing pros and cons in every direction. Something in each of them craved a little fame and notoriety. It's a good thing to be recognized nationally for something you've done. Maybe it strokes the ego in just the right way. It's a cool way to show that grumpy old math teacher from the 5th grade that you weren't a total loss. "See! You made something of yourself after all!"

Then too $50,000 each was a really nice chunk of change. They talked and talked. Finally it was decided; they would do the one show and that was all they would do. They still had a lot of work on the drawing boards. They did not want to join the celebrity circuit, even though they might make several years wages in three months that way.

The next day, Denise called Ellen's office and they agreed to do the show in one week. The sponsors of the show offered to fly them to Burbank California and back. They would stay at the luxurious Hyatt Regency Hotel.

Jim would prefer to take the Learjet. Money wasn't the issue. He just liked being in the privacy of his own jet, rather than flying commercial. So, they agreed to be there for the show at a special time. Jim would fly them out there.

They did the show and they collected the money. The team actually discovered that it was a fun experience to do the Internationally televised show. Ellen was a blast and so easy to talk to. But, now they were becoming famous and they had no desire for fame. It was the only television show they ever did. They did not wish to become household names.

Soon, it was time to go commercial with the SL-a1 model of the reactor. As a tongue-in-cheek surprise for the NSA, they built and placed their first copy of the reactor on the grounds of the NSA for free. It was tongue-in-cheek because the NSA was the largest consumer of electricity within the entire state of Maryland. Of course that meant they had always been Maryland's largest consumer of fossil fuels as well.

To offer the free reactor to the sister agency of the DOD seemed fitting since there had been such a staunch faction of big-oil, residing within their walls. Those individuals were no longer within the NSA, though. A few of those people were now languishing at the USP Leavenworth, (Federal prison). They had simply disappeared as a part of the investigations conducted by the DOD.

One cold afternoon in November 2018, the SL-a1 installation at the NSA was complete. A special ceremonial switch was thrown and the reactor was fired. It whirred to life and began producing all the **ECSAFE**: **E**conomical, **C**lean, **S**afe and **A**bundant **F**ree **E**nergy that the NSA would ever need. They were thrilled.

On completion of the Starshine Project, the team returned to Washington DC. The team members all stayed together and obtained funding towards their joint enterprise. They built a factory in the warehouse district of Washington DC and hired hundreds of workers to mass produce copies of the compact SL-a1 reactor. It could easily be toted anywhere on a flatbed tractor-trailer rig. Of course the U.S. government would retain a healthy percentage of the profits, since they had funded the Starshine Project in the beginning, but the ample profits were spread to the employees and the directors of the new factory they called the SRC (the **Starshine Reactor Corporation**).

The SL-a1 wasn't cheap. Each unit sold for approximately $8 million dollars, but cities and municipalities everywhere in the World began the switch to fusion electric. The $8 million price tag per unit seemed trivial when compared to what had previously been allotted for fuel alone. Switching to fusion electric was more than cost effective. It was 100 times more economical and efficient than fossil fuel energy.

As time passed the Starshine Reactor Corporation (SRC) continued experiments and eventually reduced the size of the reactors to almost half of the original. The newer units were named SL-a2, SL-a3 and they would eventually go all the way to SL-a10. The smaller units sold for under $1 million dollars. The nickname 'Slave' stuck and applied to all models. Out of the way hamlets and small villages everywhere in the World could now afford to have electrical power. Many of those places had never enjoyed so much as an electric light bulb before the Slave came along. Now, everyone had electric power.

The World went ballistic for fusion from that point on. Planes, boats, buses, transport, space exploration all made the switch to fusion heat and fusion electric. NASA began using the small reactors in space probes and eventually came-up with an interstellar fusion propulsion engine that would carry mankind to the outer Solar System and back for a few hundred dollars and 30 or 40 pounds of fuel.

Gas and oil stocks plummeted. We no longer needed the World's oil reserves. As a result the DOD began closing all their bases on Arab soil. Oddly enough, for some inexplicable reason, the terrorists were very pleased with that idea! They immediately stopped being the crazy lunatics they had always been. They saw it as a major win for them, and Islam. They declared that the

Jihad was officially over. The Terror Wars faded into history. We all became friends again. (No one could understand the sudden change in their attitudes!)

The tourism industries in all countries were soon booming. World economies soared to unprecedented levels. Starvation and poverty were at all-time lows. Everyone had jobs and enough money to live very well.

The Gulf of Mexico slowly became as crystal clear as the Caribbean. It actually contained fish. The polar ice caps stopped melting and it actually rained twice in southern California in the same year!

Kids went back to cruising main street in their souped-up electric hot rods, almost like the 1950s, except it was quieter. Drive-In movies and Car Hops in short-shorts returned, because people could afford to cruise and have fun in the new electric society.

New autos were now going 1000 miles on $2 worth of electricity. Freeways and Interstate hi-ways became whisper quiet in the safe, bumper to bumper auto-drive traffic. Drag Races and dragsters still existed but they too where whisper quiet! Nascar survived. The need for speed survived but it was all electric. Some of the thrills of gas combustion noise, were lost forever. Everyone missed that, but the racing spirit survived.

In July of 2018 Denise and Jim tied the knot at a double wedding with Audrey and Manuel Ramos Rodriquez (aka; Enrique T. Smith). Gus met three adorable, bright young girls and settled down with all three of them to some degree. David was still playing the field.

Hennrietta Clayton served two terms as the first woman President. General Frank Wilson retired from the DOD at age 58, but as a Democrat he threw his hat in the ring in 2024 and became President of the United States.

And . . . the World became a better place because of a few bold people . . . who dared to dream.

The End

About The Author

W.E. (Bill) Powelson hails from the deep South Texas border town of McAllen. He is currently retired (err-uhh, unemployable) and living happily in Daytona Beach, Florida, on the twelve dollars and fifteen cents he saved as a working Honky Tonk drummer.

After a lifelong career playing music (drums) for his supper, he is now in his golden years and has discovered that writing (for the fun of it) helps to keep a smile on his face. He is the author of five digital (html) e-books on the art of drumming; all of which may be viewed and studied online (or off), by going to "The Homestudy Institute of Drums" on the World Wide Web.

Other Fictional Novels by W. E. Powelson:

1. Winds of Infinity: Accelerating Earth's Technology 3000 Years.
Sci-Fi Futuristic: A middle-aged MIT professor proves Super-String and "M" theory in a unique and novel way; thus unifying all the forces of nature. In so doing, he and his soul-mate Molly also discover more than they had anticipated as they unexpectedly help an entity they call UnK bring all the religions of the world together under one unified roof . . . at the head of science.

2. Buzzard Bait: The Adventures of EZ Zeke McBride
Southwestern Fiction: A young 49er prospector, Zeke McBride is dying of thirst and starvation in the South Texas desert. He reverses his luck with the twist of a knife and then finds himself on an adventure that leads to a lovely soul mate and a pile of gold. Together, along with a host of great friends they meet along the way; they show the Wild East how good fortune and a few miracles can happen in the Wild West.

3. Tornado Gold:

Family Fiction: In the early 1990's a single mom and her two teen aged boys find themselves stranded and broke on a nowhere farm in the Texas Panhandle. A disastrous tornado takes aim at their dot on the map called South Draw and becomes a blessing rather than a calamity. It causes an unexpected turn of events which puts a finish to the tale of (Buzzard Bait) Zeke McBride's lost gold.

A family adventure of Disney proportions awaits them as one unanticipated surprise follows another.

4. Granny & Pooch VS the Garbage Mafia'

Fiction CE: The howling Doberman next door drives Granny up a wall late at night. She visits with a peace-token-treat of left-over roast beef. As a result she not only inherits Pooch the Doberman, but she discovers a murder and solves it as she almost becomes a victim herself. Along with the mayhem, she also uncovers a surprise windfall of adventures; including potential answers to what became of the Mafia. It's a fictional story. (. . . Or is it? You decide!)

5. The Irrelevant Few: "Redeemed"

Futuristic fiction, 2034 CE. Homeless populations have exploded. Congress acts to house and feed the indigent hoards, but the initial plan goes awry. Incarceration becomes mandatory for those without dependable incomes. Additional pressures are meted by secret leaders within the Pentagon. An *irrelevant few* struggle to right the wrongs and regain eroded freedoms for the indigent multitudes. It could happen . . . SOON!

6 Fusion!

A cracker-jack team of four MIT grads get selected by the Research and Development wing of the DOD to develop sustained fusion; (the obvious choice for energy of the future). The government of the United States views their efforts as an extension of the Manhattan Project that produced the Atomic Age. It is that important! A story of love, mystery and intrigue evolve as the Starshine Project team attempts to get the job done within the laboratories of Los Alamos, New Mexico.

7. E-Book and Print On Demand Publishing

Smashwords Formatting -(Free e-book))
"Open Office Through the Meatgrinder".
Non-Fiction: This free e-book is my own zany attempt at unraveling the mysteries of Smashwords e-book formatting. As one who was too cheap and too broke to pay for Microsoft's "Word" word processing

software; I was forced by the economics of the situation to use the free version of Open Office.

At this point I now recommend using Open Office all the way. The truth is, I made some discoveries that condense the job down to about a one-hour learning-curve for most people. The trick is to use a Template as you begin. The free templates I've designed will set the internal gears of the software AUTOMATICALLY to Smashwords standards. After that, it becomes a very simple matter to use almost ANY full featured word processor to format most books in an hour or two.

I'm actually being a little too modest. You will shave approximately two weeks off the learning curve by using my FREE templates and the (also free) short 4-chapter guide I wrote. The guide and the templates are totally FREE at the following web address (below).

HOW TO:

Read the web page below; download the free template you'll need, and a free copy of "Open Office Through the Meatgrinder". It's all free. Your formatted e-book may be finished by the time you reach chapter 3 of my guide.

Yes! It's that easy! Click the link below to get started.

"Open Office Through The Meatgrinder":
<http://www.studydrums.com/templates/index.html>

PUBLISHING HELP:

If you are seeking help with Smashwords Meatgrinder E-book Formatting or Create Space '**Print On Demand**' formatting; e-mail Bill at: W.E.(Bill) Powelson <billp@studydrums.com>. Bill will be happy to help in every way he can, free of charge, plus (if you prefer) he will format your e-books according to Smashwords (or Create Space) specs for a fee of $25 per each 100 pages.

Together we'll get it done!